The Deep Blue Planet

BACK TO THE SEA

The Deep Blue Planet

BACK TO THE SEA

RENATO MASSA
ENGLISH TRANSLATION BY NEIL FRAZER DAVENPORT

RAINTREE
STECK-VAUGHN
PUBLISHERS
The Steck-Vaughn Company

Austin, Texas

Published by Raintree Steck-Vaughn Publishers, an imprint of Steck-Vaughn Company

Editors
Caterina Longanesi, Linda Zierdt-Warshaw, William P. Mara

Design and layout
Jaca Book Design Office

Library of Congress Cataloging-in-Publication Data

Massa, Renato.
[Ritorno al mare. English]
Back to the sea / Renato Massa ; English translation by Neil Frazer Davenport.
p. cm. — (The deep blue planet)
Includes bibliographical references (p. 50) and index.
Summary: Describes a variety of animals that live in the sea, including sea turtles, penguins, and whales.
ISBN 0-8172-4653-3
1. Marine animals — Juvenile literature. 2. Cetacea — Juvenile literature. [1. Marine animals.] I. Title. II. Series.
QL122.2.M37513 1998
591.77 — dc21
97–12102
CIP AC

Printed and bound in the United States
1 2 3 4 5 6 7 8 9 0 WO 01 00 99 98 97

CONTENTS

Please note: words in **bold** can also be found in the glossary. They are bold only the first time they appear in the main body of the text.

INTRODUCTION

The oceans and seas are wondrous places of both dark and light, flora and fauna, warm and cold, and life and death. Many regions are still largely unknown to us, so there is little doubt that hundreds of mysteries still lie in their murky depths.

The oceans and seas cover just over 70 percent of the Earth. They have provided us with many pieces to the puzzle of how life began on this planet, since the earliest forms of life were thought to have existed there. The rocks and sediment along the ocean floor have yielded a great deal of fossil evidence over the years.

The value of the oceans and seas is immeasurable. They have given us many useful chemicals and minerals, including bromine, magnesium, and salt, not to mention pearls for jewelry and shells for building material and health supplements. Most experts believe we have not yet realized their full potential in regard to nutrition, though it is believed that humans derive at least 10% of their overall protein from the Earth's waters, either directly or indirectly. Finally, there is the recreational aspect. Activities such as swimming, fishing, boating, diving, and so on, when executed properly and responsibly, provide us with a great deal of pleasure and a measure of relief from the grind of our daily lives.

Sadly, however, we humans have caused some serious damage to the oceans and seas in recent times. Industry is the greatest violator, with over a quarter of a million manufacturing facilities using the great bodies of water as dumping grounds for their often highly dangerous waste products, including mercury, lead, sulfuric acid, and asbestos. In addition, towns and cities regularly dump improperly treated sewage and millions of tons of paper and plastic wastes into rivers, streams, and lakes. Plastics in particular have the potential to remain intact for hundreds of years.

However, we have not yet reached a point of no return, and one of the goals of the *Deep Blue Planet* books is to give you a deeper understanding of—and in turn a deeper appreciation and respect for—the aquatic environments of this world. The more you know about any subject, the greater your appreciation for it will be, and the oceans and seas are in desperate need of increased appreciation. Perhaps someday you will make efforts of your own to preserve these beautiful natural areas and the myriad life forms that thrive within them. If so, you will be helping to guarantee them the bright and vibrant future they so richly deserve.

Back to the Sea

Recurrent Nostalgia

According to the most commonly accepted theory, all living creatures on Earth originated in the oceans and seas. **Vertebrates**—animals with backbones—originated from ancient life forms similar to modern lancelets and the larvae of tunicates. Vertebrates remained aquatic for hundreds of millions of years, gradually evolving into jawless fishes, armored fishes, cartilaginous fishes, and bony fishes.

In time these early fishes populated the world's fresh water as well as the salt water. They finally appeared on land during the **Devonian** Period. The first of these creatures were amphibians, and they still needed to reproduce in water. Later on, **reptiles**, who could lay eggs that remained fertile in dry environments, evolved.

From the reptiles onward, the freedom of land animals from the seas was complete. Vertebrates moved further onto the dry terrain. They first evolved as reptiles and **birds**, and then the **mammals** evolved. The **adaptations** of the great vertebrate classes and the early insects ensured that they could all survive while being strictly **terrestrial**.

There is a difference between the evolution of the vertebrates and the insects—the insects have never gone back to the sea. This may have occurred because the sea contained other arthropods, like the many crustaceans and a few arachnids (e.g., horseshoe crabs and sea spiders) that were better adapted. Land-based vertebrates, however, have made many return trips to the sea. It's almost as if they were driven by an evolutionary nostalgia for the environment from which they came.

Dozens of reptile species live in today's seas. They belong to all the living orders except the Rhynchocephalia (which contains only one creature, the lizard-like tuatara). More than 100 species of mammals, belonging to four distinct groups, also live in the seas. About 200 bird species, from four orders, also make the sea their home.

The Golden Age of Marine Reptiles

At the height of their glory—in the middle of the Mesozoic Period—there were many marine reptiles. Above all were the **ichthyosaurs**. These amazing animals are similar to modern dolphins. The ichthyosaurs share the dolphins' jaw shape, swimming ability, and vivaparous (live-birth) reproductive system. This last trait was shown by **fossil** remains of a mother, possibly buried by falling volcanic ash, as she was giving birth. The offspring emerged tail first, just as they do with modern dolphins.

The evolutionary meeting place of ichthyosaurs and dolphins came from a total adaptation to a marine lifestyle. Like dolphins, the ichthyosaurs could not give birth on dry land. They first appeared during the **Jurassic** and were extinct by the upper **Cretaceous**, many millions of years before the end of the dinosaurs.

Another impressive group of marine reptiles were the **Sauropterygia**. These animals were sea dragons. They had bodies almost 15 meters (49 ft) long. They had curving and sometimes long, snake-like necks. Their mouths had frightening jaws filled with sharp teeth.

Sauropterygians were present throughout the Mesozoic Era, following two major evolutionary changes. One form of **plesiosaur** (for example, *Hydrotherosaurus*) had a long neck and shortened fins. Another form (for example, *Cryptocleidus*) had a shorter neck, a larger head, and more hardy fins. This second form was the oldest. It appeared in the **Triassic** Period along with other marine reptiles that were less specialized and more similar to modern finned mammals, such as the **nothosaurs** and the placodonts.

As mysteriously as the dinosaurs, the Sauropterygia disappeared from the Earth. They left no trace except for a few fossils that were found and studied by paleontologists. Scientists have also discovered the remains of giant marine turtles that had shells 4 to 5 meters (13 to 16 ft) long.

1

3

1. A superb example of a fossilized ichthyosaur, *Stenopterigius quadricissus*. This fossil was found at Holzmaden (Germany) and dates back to the Jurassic Period. The origins and reasons for the disappearance of the ichthyosaurs are not known. **2.** Ichthyosaurs in their native marine environment. The ichthyosaurs were similar to modern dolphins. They were between 2 and 15 meters (6.5 and 45 ft) long. They were also carnivorous and viviparous. Ichthyosaurs lived in the seas of the Mesozoic Era, from around 200 to 65 million years ago. **3. 5.** *Neusticosaurus peyeri* was a nothosaur. This reptile group was well adapted to an aquatic environment, although less so than the ichthyosaurs. **4.** *Nothosaurus, Elasmosaurus*, and *Peloneutes*. The first is a nothosaur. The other two are plesiosaurs, which were aquatic reptiles of the Mesozoic Era.

2

4

5

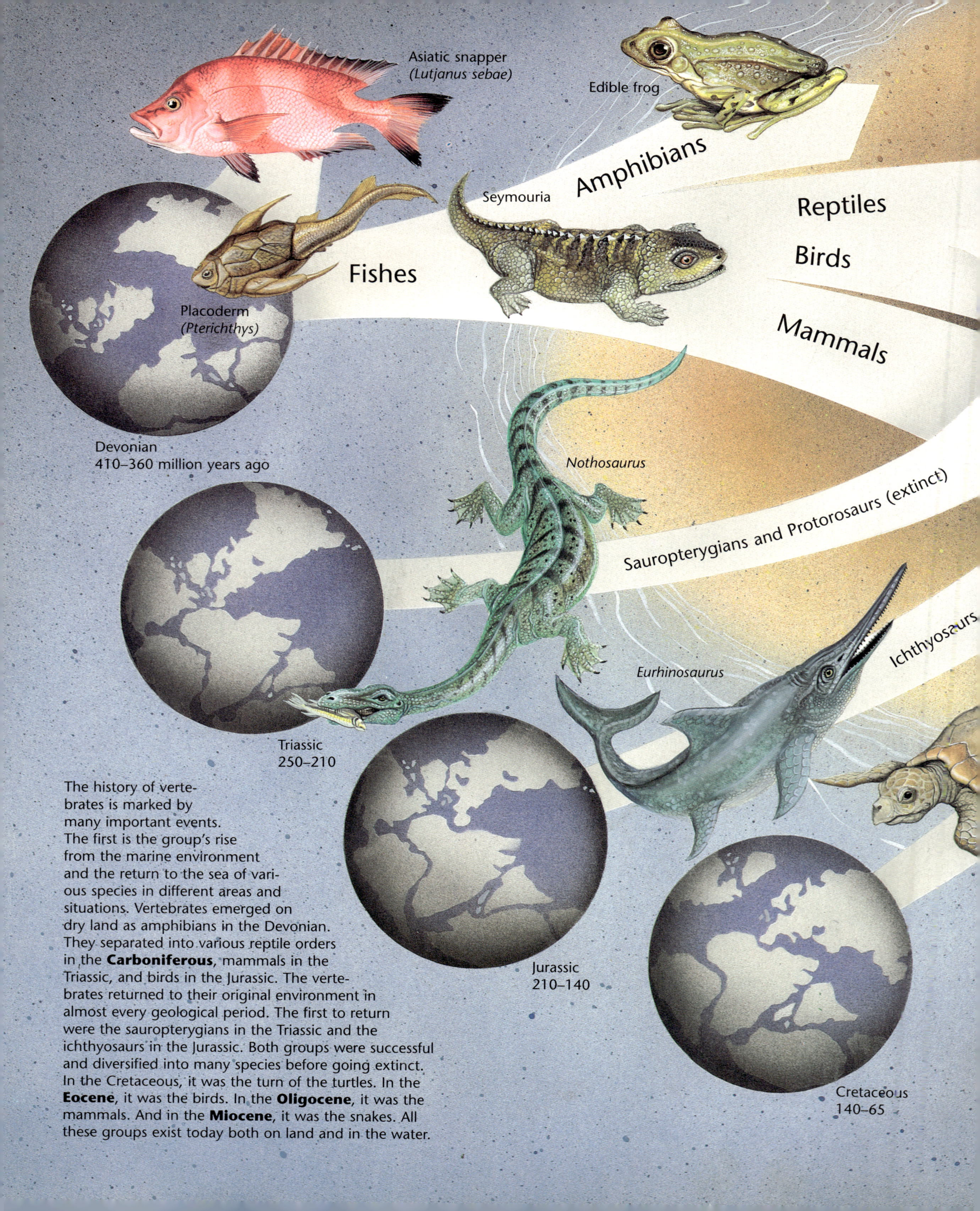

The history of vertebrates is marked by many important events. The first is the group's rise from the marine environment and the return to the sea of various species in different areas and situations. Vertebrates emerged on dry land as amphibians in the Devonian. They separated into various reptile orders in the **Carboniferous**, mammals in the Triassic, and birds in the Jurassic. The vertebrates returned to their original environment in almost every geological period. The first to return were the sauropterygians in the Triassic and the ichthyosaurs in the Jurassic. Both groups were successful and diversified into many species before going extinct. In the Cretaceous, it was the turn of the turtles. In the **Eocene**, it was the birds. In the **Oligocene**, it was the mammals. And in the **Miocene**, it was the snakes. All these groups exist today both on land and in the water.

Nile
crocodile
Reptiles
Tawny
owl
Birds
Testudinidates
Emperor penguin
Birds
Leopard
Mammals
Loggerhead
turtle
Sea snake
Blue whale
Eocene
65–37
Oligocene
37–23
Miocene
23–5

Sea Turtles

Testimony to a Glorious Past

Among the great marine reptiles of the past, the oldest, most common, and possibly most important group that still exists is the sea turtles. There are just five species in this group, and they live throughout tropical and temperate waters from North to South America, through to Eurasia, Africa, and Oceania.

Sea turtles are divided into two families. The best-known is the Chelonidae. This family includes four species with a

1

"classic" sea-turtle structure—the loggerhead (*Caretta caretta*), the green turtle (*Chelonia mydas*), the hawksbill (*Eretmochelys imbricata*), and the olive Ridley turtle (*Lepidochelys olivacea*). The second family is the Dermochelidae. It contains one species, of a remarkable size and unusual appearance—the leatherback (*Dermochelys coriacea*).

The four Chelonidae species have a bony shell. The shell is made up of a **carapace** and a **plastron** formed from large plates that are bonded together. The leatherback has the same shape as other sea turtles, but, at first sight, it appears to lack a shell. It instead seems to be covered by a smooth, streamlined skin with seven deep channels running the length of the body. In reality, a bony shell does exist, but it is made up of small bony plates that are hidden beneath the skin. This uncommon turtle cannot withdraw into what you might call its "virtual" shell. However, the turtle makes up for this disadvantage with its huge size. Leatherbacks may be more than 2 meters (6.5 ft) long and weigh over 500 kilograms (1,102 lbs). Today, examples weighing more than 350 kilograms (772 lbs) are rare. The span of its outstretched front fins may be as great as 270 centimeters (686 in). The leatherback is carnivorous. It feeds on fishes, crustaceans, octopuses, and jellyfish. It also is a fast swimmer, although not for extended periods of time.

Nesting Turtles

The adaptation of the Chelonidae family to the marine environment is less extreme than that of the Dermochelidae. The Chelonidae family has retained the basic structure common to all turtles. It also has a link with dry land. For example, these species often approach the coasts to feed on fish in shallow waters.

Notably, all sea turtles return to dry land to complete their reproductive cycle. They undertake migrations that often involve traveling great distances in order to reach the warm beaches of the tropics so they can lay their eggs. To this end, the gravid females must leave the water and move onto dry land above the high-tide mark. There they dig holes with their rear limbs and lay piles of eggs that sometimes number in the hundreds. This task is usually carried out at night and takes several hours.

The long time required by the egg-laying process of sea turtles is due to their great weight (up to 200 kilograms [441 lbs] in the case of loggerheads and almost twice that for green turtles) and the awkwardness of trying to use their limbs, which are adapted to sea life, on land.

Digging into the sand with their front and rear flippers, they drag themselves forward a short distance before stopping and gasping deeply. They become short of breath because they are not suited to moving on dry land and because of the effort needed to lift their heavy bodies and allow their lungs to fill with air. If a mother sea turtle remained on dry land for longer than necessary to lay her eggs, she would suffocate from her own bulk. When the mothers finish laying their eggs, they throw the sand back in all directions, sloppily trying to erase evidence of their activities.

1. A loggerhead turtle and a diver in the Red Sea. **2.** A view of a loggerhead from below. **3.** Leatherback turtle (*Dermochelys coriacea*). **4.** Green turtle (*Chelonia mydas*). **5.** Loggerhead turtle (*Caretta caretta*). **6.** Pacific green turtle (*Chelonia agassizi*). **7.** Hawksbill turtle (*Eretmochelys imbricata*). **8.** Olive Ridley turtle (*Lepidochelys olivacea*). **Following pages:** Female green turtles have laid their eggs during the night on a tropical island beach. They are now returning to the sea. The males—recognizable by their large, long tails—are waiting in the water just off the beach. They will meet the females as they complete their land journey and mate with them to ensure the next batch of eggs is fertilized.

8

2

7

6

5

4

3

DEL NEVO A.

SEA SNAKES

1 2

Small and Venomous

Another major group of reptiles with many marine species is the snakes. Sea snakes are not very large—only 1 or 2 meters (3 to 6.5 ft) in length—but they often are quite attractive, some with bright colors and well-defined patterns. The colors warn predators that they can be dangerous, and indeed they are, for sea snakes are highly venomous. They are classified in a family of their own—the Hydrophiidae—but they are very closely related to the cobras (of the family Elapidae). Technically, sea snakes are proteroglyphs, which means they have small and fixed front fangs. These fangs channel **venom** from venom ducts located in the roof of the mouth and are directly attached to the fangs.

The venom of most sea snakes is of the neurotoxic type, like that of cobras and scorpions. Its potency varies according to the species and their individual needs. For example, in seas that are rough and have uneven floors, the venom of the native sea snakes is often very powerful. Powerful venom is needed because the prey—usually an eel or a fish—needs to be immobilized before it can move away and die in an inaccessible recess. In areas where the sea floor is smooth and the water is clear, the snakes' venom is less powerful. Here the prey is less likely to move out of the snake's field of vision.

Open-Sea Swimmers

All sea snakes are well-adapted to their environment. Sea snakes have a serpentine form characteristic of all ophidians (snakes). This form is well-suited to swimming. Sea snakes also have small heads. Their necks and the front parts of their bodies are fairly slim. The rear parts are thicker and compressed laterally to form a kind of "rudder" that helps them steer. A sea snake's nostrils are located on the upper part of its snout. It uses these to breathe when it comes to the water's surface. Contrary to what some people may think, sea snakes are not able to respire underwater the way fish do.

The shape of the **scales** of sea snakes separates the two subfamilies. The most primitive is the Laticaudinae. Members of this group have large ventral scales in a single series, like those of land snakes. In contrast, the Hydrophiinae have small, overlapping ventral scales like those on the backs of most snakes. Thus, the entire body has a smooth, streamlined shape that is good for swimming but not for moving on land. Most sea snakes give birth to live young and therefore do not need to return to land to lay eggs. The only exceptions are certain Laticaudinae species.

Sea snakes are most common in Australasian waters. From the Persian Gulf and the waters of the Indian Ocean, they stretch as far as southern China, Taiwan, Indonesia, New Zealand, Australia, and near several Oceanic islands. Sea snakes are even fairly common in some of these areas and are well-known among the native peoples. However, they are not particularly feared because most species are not aggressive to humans. They bite only when intensely provoked. Nevertheless, it should be pointed out that sea snakes possess the most potent venom of all the snakes in the world.

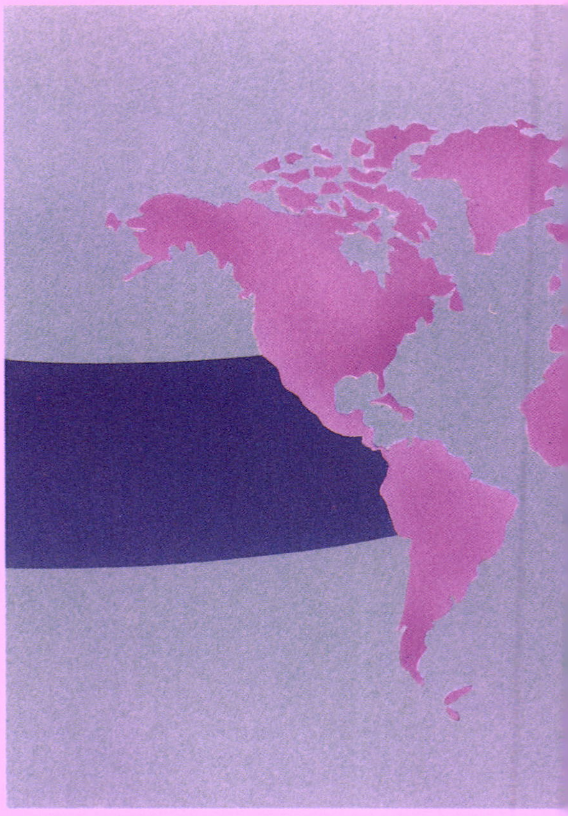

Preceding page:
1. A sea snake with a broad, flat tail. *Laticauda colubrinus,* the yellow-lipped sea snake, with its eggs on a pebble beach. This beautifully colored species retains traces of its past as a land reptile. It returns to the shore to lay its eggs and to warm itself in the sun. In contrast, many other species in the Hydrophiidae family are completely adapted to the marine environment. They give birth to fully developed young right in the water. **2.** The white-bellied sea eagle, *Halaetus leucogaster*, snatching a sea snake from the water.

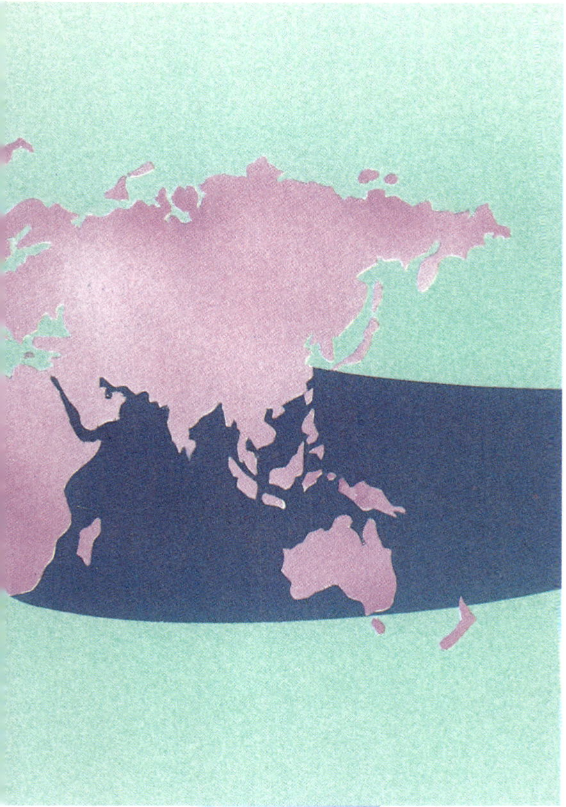

1

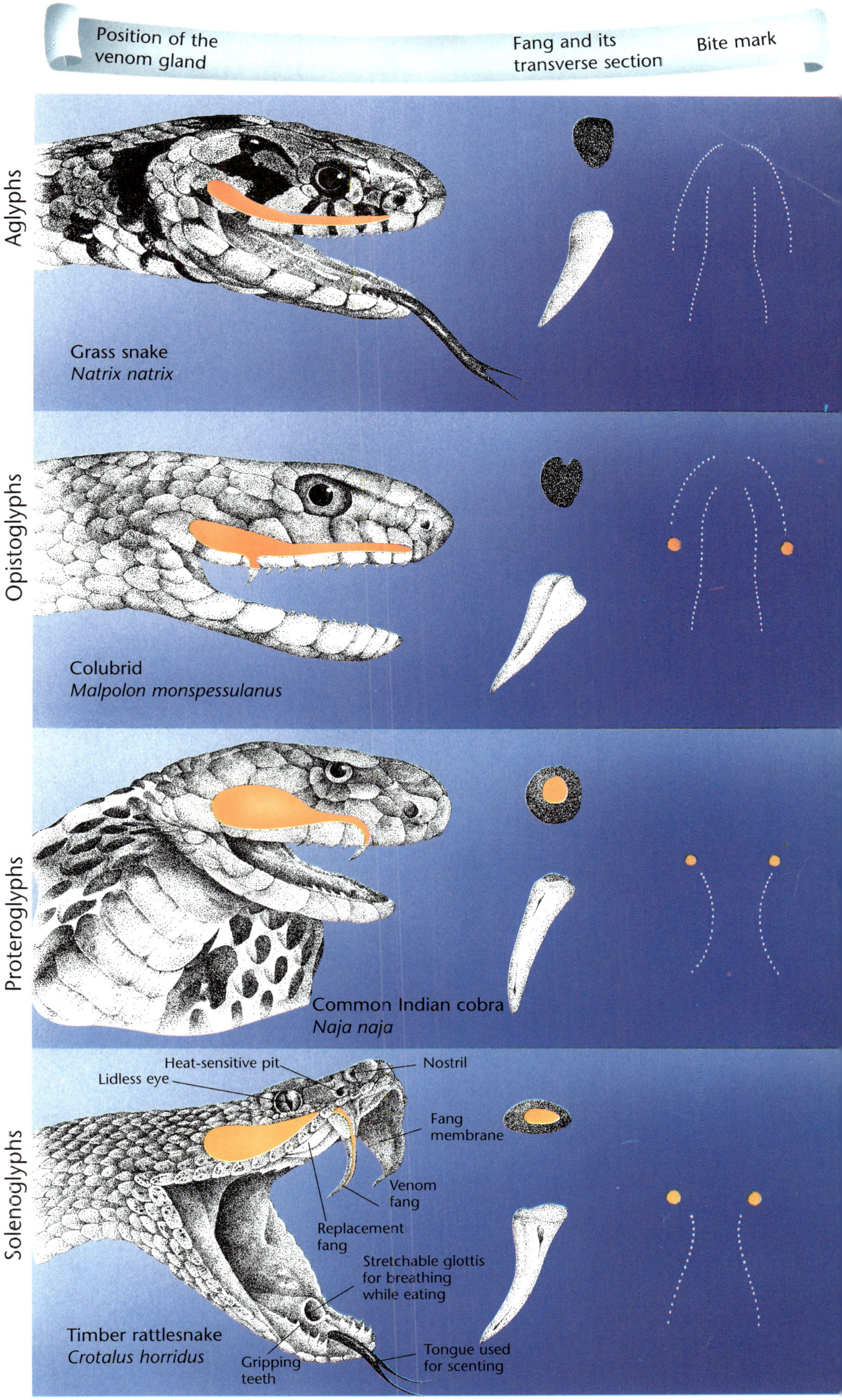

3

1. The geographical distribution of the sea snakes that form the Hydrophiinae family. **2.** The small island of Pulau Kalampunian Damit is part of the Pulau Tiga park off the coast of Sabah, Borneo. Known for the frequent presence of Hydrophiinae, it has been nicknamed "Snake Island." **3.** Types of teeth and bite marks of the snakes living today. The aglyphs have solid teeth that are all identical and have no venom ducts or fangs. The opistoglyphs have two fangs that are solid and channeled but are set back in the jaw. The proteroglyphs and the solenoglyphs have two front fangs with internal ducts through which the venom flows. In the solenoglyphs, the venom fangs can be folded up.

Winged Fishers

The Conquest of the Water-Air Boundary

The modern class of reptiles has colonized the open sea with sea turtles and sea snakes, and the coastal environments with lizards and crocodiles. The modern class of birds has been equally adventurous. For the conquerors of the air that appeared in the Jurassic Period and today number over 9,000 species, movement toward the sea must have resulted from strong evolutionary pressure. This movement was not a true return (as with sea turtles and **cetaceans**), but rather the possession of a new ecological niche between air and water.

Sea birds belong to at least four groups. The first is the procellariiformes and relatives that include the various **Procellariidae**, shearwaters, fulmars, petrels, albatrosses, and storm-petrels. All these species are highly adapted to a **pelagic** environment. The second group is the pelecaniformes and their relatives, including pelicans, **gannets**, **cormorants**, **frigates**, and **bos'n birds**. These species are coastal birds that depend on the sea for food. The third group of sea birds is the charadriiformes and relatives. They include waders and gulls—birds mainly tied to the coasts—and **auks**, which are as pelagic as the procellariiformes. The fourth is the **penguins**, a group of flightless birds greatly modified for the marine environment.

The modern classification of birds by Sibley, Ahlquist, and Monroe is based on DNA rather than structure. In this classification, all seabirds belong to the order Ciconiiformes. This order includes at least 29 families. The families are classified in various super-families and in other transitional groupings between order and family. In this and the following chapters, we will deal only with pelagic seabirds, the procellariiformes, auks, and penguins.

Angels of the Sea

The procellariiformes and relatives (all in the Procellariidae family) are distinguished by nostrils formed from horny tubes that run along the top of the beak. A unique desalination system occurs within these nostrils, allowing members of the Procellariidae to drink sea water. The group, made up of the albatrosses (the Diomedeinae subfamily), contains the largest species. Some albatrosses have wingspans of up to 3.5 meters (11.5 ft).

Four different phases in the complex courtship dance of the albatrosses. The dance is accompanied by shrill cries and a clapping of beaks. The whole albatross family practices strict and perennial monogamy, a rare social structure in the animal kingdom. The motives favoring the development of this faithfulness are very practical—the need to organize long and arduous shifts in order to feed the offspring for many months, and fishing far out to sea before returning to the nest. Through the courtship dance, the two birds express reciprocal solidarity and achieve a perfect understanding of one another. These psychological states are important when the birds have to tackle the hard work of raising their young.

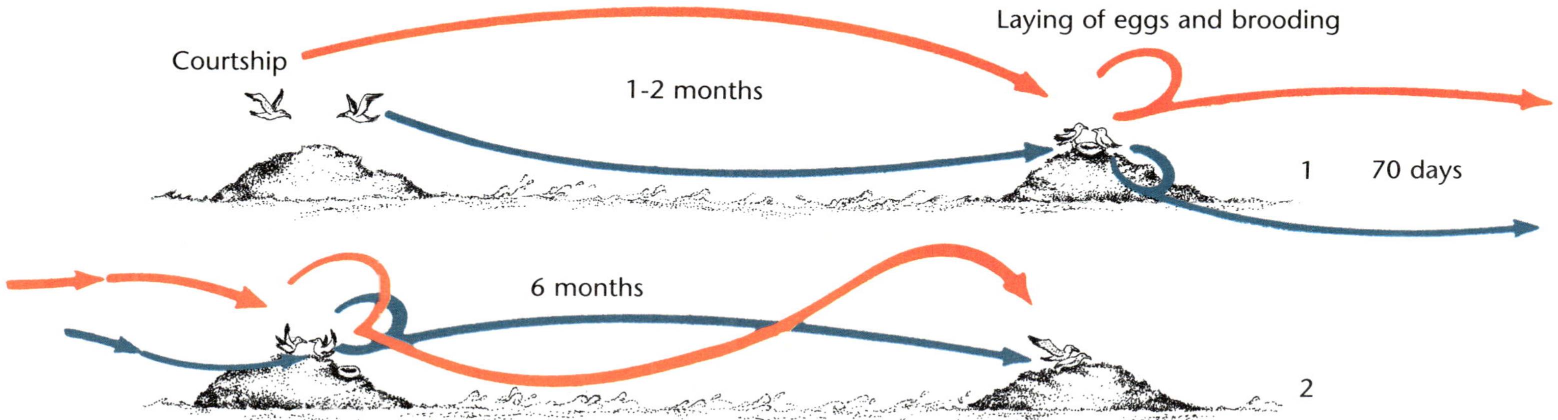

These birds are graceful and efficient fliers. They live between the water and the air, touching down on the ocean surface to fish and to sleep. They remain at sea for months and return to land only to build their nests on the coasts of small islands.

Albatrosses are sexually mature at four or five years of age. They enter into a "marriage," to which they are completely faithful. The marriage begins with a complex courtship dance that must be practiced many times before it can be done perfectly. The ritual is difficult but necessary if the albatrosses are to reach the psychological state needed for coupling.

In their fourth year, albatrosses land at the traditional breeding stations. Once they find a suitable partner and have succeeded in performing the courtship dance, the "fiancees" do not then mate. Instead they part and return to sea to spend a lonely winter on their own. But they meet up again the next spring, at the same breeding station, and then mate.

After breeding, the male and female take turns brooding the single egg and feeding at sea, often hundreds of kilometers away and for many weeks. When the egg hatches, the pair begin a frantic shuttle back and forth to feed their offspring. During this time they work alone but for the same purpose. Only rarely do they meet at the nest. Once the young bird is weaned, the parents take to the seas, alone again, but with an appointment to meet the following year.

1. 2. The inseparably paired albatrosses spend only a few days together each year. This happens when they return to the small island in the ocean on which they build their nest. After the courtship ceremonies and mating, the pair take turns brooding and hunting. When the young bird has been raised, the adults go their own way until the next mating season.

3

3. Waved albatross, *Diomedea irrorata*. Most of the 13 species of albatrosses live in the cold oceans of the Southern Hemisphere. Some species reach the tropical zones of the Pacific. For example, the waved albatross nests in the Galápagos Islands.

1

Predators and Prey of the Sea

Albatrosses feed mostly on fishes and squid. The same is true of shearwaters and fulmars (of the Procellariinae subfamily), which are smaller and more widely distributed variants. The fulmars of the **Arctic** and **Antarctic** oceans have adapted to feeding at least partially on the waste products of the human fishing industry. As a result they have increased greatly in number over the last century. In contrast, the very small "storm petrels" (many species weigh no more than 25-30 grams [0.9-1.1 oz], about the same as a sparrow) feed mainly on zooplankton, especially small crustaceans that they gather from the sea's surface. Given the abundance of this resource, the numbers of storm petrels is also very high.

The smaller Procellariidae species nest in tunnels—which sometimes are abandoned mammal burrows—to avoid predators, especially large gulls. On land they are awkward and helpless. This is due to the fact that their legs are set very far back, forcing them to adopt an uncertain, rocking gait. Shearwaters enter and leave their nests only at night. They are sometimes found floating on the water in large groups, near sunset, as they wait for the cover of darkness so they can escape the clutches of the fierce winged pirates that threaten them. Shearwaters have a strong, musty odor. They probably use their sense of smell to find their nests and their partners in darkness.

Penguins of the North

The other large group of pelagic seabirds includes the auks, razorbills, guillemots, puffins, and their relatives. Like the Procellariidae, this is a highly specialized group. They live much of their life in the water rather than in the air.

All auks have small, strong wings, which are excellent for maneuvering in tight spaces. The wings are also used as fins when swimming underwater. They are unsuitable, however, for gliding and for long-distance flights like those completed by the Procellariidae family.

Auks resemble penguins in general appearance. They have a very erect posture and often have black plumage on their backs and white underparts. The great auk, hunted to extinction in the middle of the last century by Danish sailors on the coast of Greenland, was flightless. This bird was actually defined as a "penguin" in its scientific name, *Pinguinus impennis*. From a functional point of view, the auks are a close northern equivalent to the penguins, which are confined to the Southern Hemisphere.

Auks nest in colonies. The smaller species (puffins, for example) occupy tunnels such as empty rabbit warrens. Outside the nesting period, all auks go to the open sea. There they are hard to spot, especially when their typical black-and-white coloring is replaced by a lighter winter plumage. While they are nesting, auks are among the most closely studied of all birds, especially by experts in animal behavior, who are always looking for subjects that are easy to locate and recognize individually. A typical example of this research is the study of the birds' ability to recognize individual eggs by the shape of the shell. This ability is particularly useful for these birds since they nest in colonies on crowded ledges where it would be easy to confuse nests and eggs if not for this innate talent.

3

2

1. A classic print by the English natural history illustrator John Gould (1804-1893) showing the dark and pale phases of the plumage of the northern fulmar, *Fulmarus glacialis*. The two variants, today generally considered to be separate species, are distributed respectively in the Antarctic and the Arctic Oceans.
2. A fulmar in flight over a Scottish island. Until 1878 northern fulmars were not found farther south than Iceland and the Faroe Islands. Over the last 100 years, however, the species has invaded the British Isles, establishing over 500 colonies with a population of about 150,000 individuals.
3. A black guillemot, *Cepphus grylle*, in winter (light) and summer (dark) plumage, shown in a print by John Gould. A more northern species than other members of the Alcidae family, the black guillemot remains in the North Sea even during the winter.

5

6

4

4. A guillemot in flight over the Firth of Forth, near Edinburgh. The guillemots have small wings and are modest fliers. However, they swim freely under water using their wings as fins.
5. A print by John Gould showing a puffin, *Fratercula arctica*, carrying food to its chick at the foot of a cliff. The puffin can keep several small fish in its beak. This helps lessen the number of trips it has to make to feed its young.
6. A colony of puffins on the rocky coast of the Farne Islands, England. At the end of the nesting period, the birds gather on the coast in groups before heading out to sea.

PENGUINS

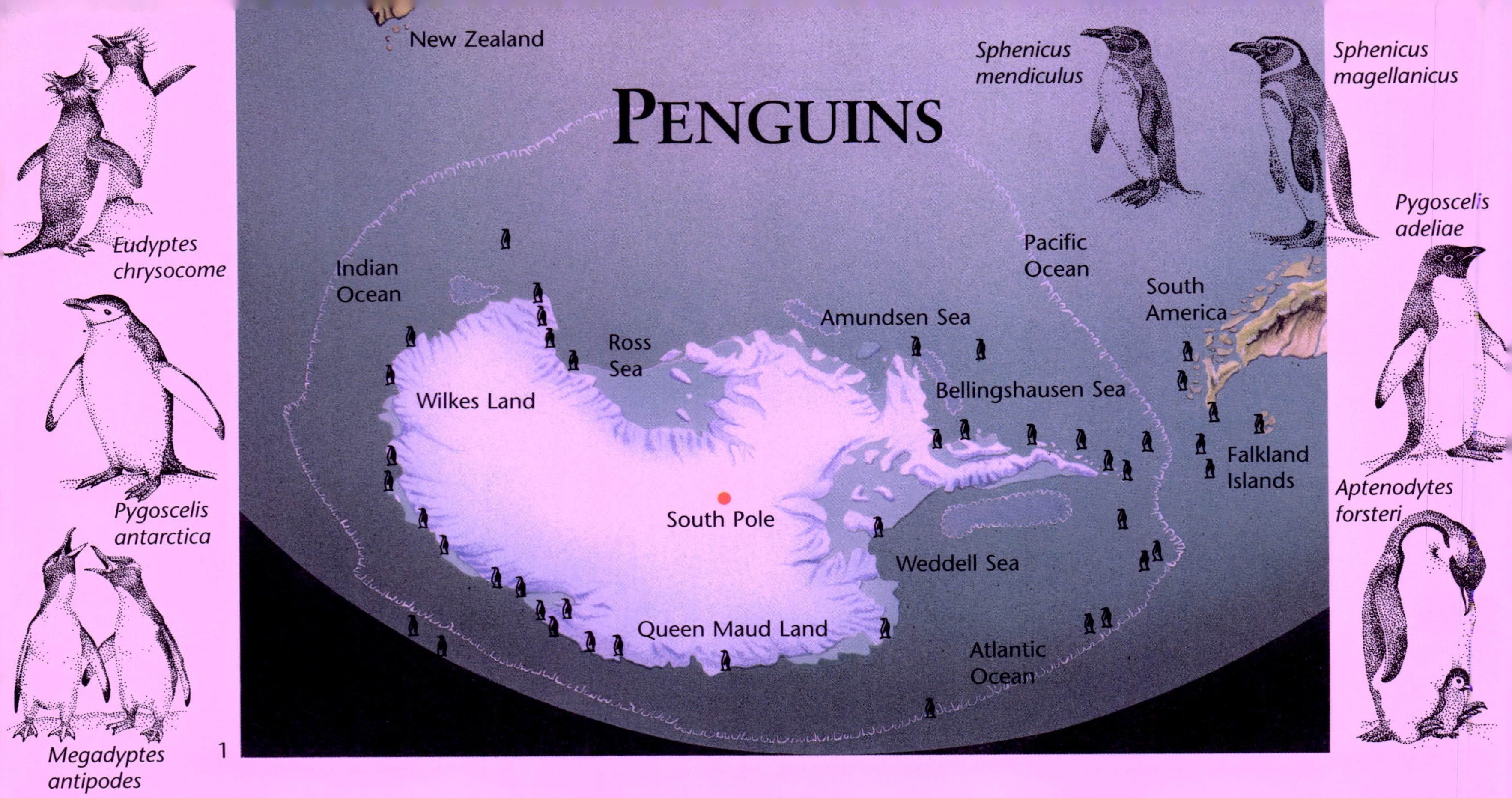

1

Extraordinary Adaptation

There are 17 species of true penguins in the **Spheniscidae** family. They are all flightless but are strong swimmers. Under the old classification, this family was placed in its own order. It was also in a separate subclass from other birds, the Impennae. Research into DNA has shown that the loss of classic feathers in penguins is fairly unimportant. It also shows that penguins are closely related to two other groups of marine birds—the frigates and the loons.

Penguins are characterized by extreme specialization. This has led to greater anatomical and physiological modification than in any other aquatic bird group.

Millions of years ago, when the ancestors of the modern penguins were still able to fly, they had to face fierce competition from other seabirds. This still happens to some extent among the procellariiformes, cormorants, gannets, gulls, and other groups. Severe competition inspired penguins to specialize in swimming and diving to great depths and for great lengths of time. The birds evolved in response to these adaptations—gradually their feet were positioned farther back, and thermal insulation outside the water was ensured by a layer of feathers covering the entire body. Also, penguins do not have any of the bare patches found in other birds; layers of fat rather than feathers keep the birds warm while swimming. The insulation system is so efficient that it creates a danger of overheating. To help lose excess heat, penguins have developed radiating surfaces that lack layers of insulating fat on the undersides of their wings and on the tops of their feet.

The Antarctic, A Penguin's Paradise

Most penguin species are found in the Antarctic. This region has many advantages for these birds. For example, it provides heat regulation and is very rich in food resources. The area supports only a limited number of potential predators. Even fewer parasites and pathogens exist because they cannot complete their life cycles in the harsh, frozen climate.

When not engaged in their reproductive cycle, penguins spend all their time in the water. Here they move with as much agility and elegance as dolphins. Penguins use their "wing paddles" to propel themselves forward. They use their webbed feet as rudders. Like dolphins, they often leap out of water, probably to breathe in and out rapidly while preserving their speed. The largest species can dive to a depth of 200 meters (656 ft) and can stay under water for up to 15 minutes.

Some penguin species complete migrations over great distances to reach their nesting grounds. The most unusual species is the emperor penguin (*Aptenodytes forsteri*), which reproduces during the Antarctic winter.

Emperor penguins move south to nest from late March to early April. They brood on the ice in temperatures of -40°C (-40°F) for about 60 days during the Antarctic autumn. They raise their young during the winter and wean them between December and January, early in the Antarctic summer.

When brooding or covering their young to keep them warm, emperor penguins must fast because their only food source is in the sea. This adaptation is so amazing that even after three months of fasting, a parent can regurgitate enough food to keep its offspring alive until its companion returns from the sea with more food.

1. The distribution of penguins is confined to the Southern Hemisphere. Two of the 17 known species nest on the Antarctic continent. Another 12 nest on sub-Antarctic islands and rocks. One species nests in South America, another on the Galápagos Islands. **2.** Adélie penguins diving from an iceberg in the Ross Sea, Antarctica. Adélie and emperor penguins are the species most adapted to extreme conditions.

2

1

2

Similar to emperor penguins but smaller and featuring an orange band on their necks, king penguins nest on Staten Island, South Georgia, the Falkland Islands, the Marion Islands, the Crozet Islands, Kerguelen, Heard Island, and Macquarie Island. **1.** An individual on parade. **2.** A colony on Salisbury Plain, South Georgia, Antarctica.

The Whale Family

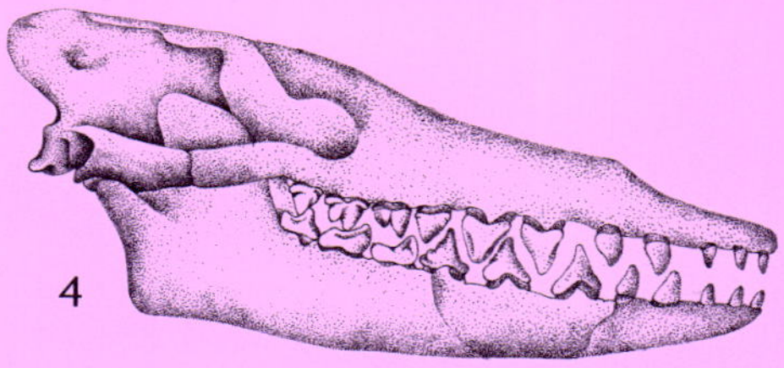

4

Origins of the Whale Family

In the story of the vertebrates' return to the sea, the most amazing chapters do not belong to modern reptiles, who have changed little to adapt themselves to life in the open sea. The most amazing chapters do not belong to birds either, who, with the exception of the penguins, have remained suspended between air and sea environments. Instead, the most extreme adaptations are those of the group of mammals known as cetaceans, or, members of the whale family.

The oldest known cetacean fossil remains were found in Pakistan. They date back to the Eocene, about 60 million years ago (a few million years after the extinction of the dinosaurs). The skulls of these ancient cetaceans resemble those of some primitive land mammals of that period, in particular the carnivore *Hapalodectes* of the family Mesonychidae. These animals were similar to the ancient land forebears of the dolphins and whales.

In the fossil remains, the snout is elongated. The dental arch contains well-differentiated teeth, with incisors, premolars, and molars. In modern cetaceans with teeth, all the teeth are identical. The teeth are used to capture prey and to chop it into large pieces.

The remains of a *Pakicetus* sp. ("sp." is the abbreviation for the singular form of "species;" "spp." is the plural) were found in a deposit that also contained many bones of land mammals. This suggests that this early cetacean spent at least part of its life on land, like today's pinnipeds (seals, sea lions, and walruses). The evolution toward a solely aquatic lifestyle must have been rapid since there are modern fossil remains that date from the Oligocene, around 20 million years later. These fossil species are now classified in the **odontocete** (dolphins, killer whales, and sperm whales) and **mysticete** (**baleen** whales) suborders.

1. The skeleton of the carnivorous mesonychid *Mesonyx*. This creature was a little over 1 meter (3.3 ft) long and lived around 60 million years ago. It is considered the most probable ancestor of all cetaceans. **2.** The skull of another mesonychid, *Harpagolestes*. The huge canine teeth and the robust bony crest onto which the jaw muscles were attached suggest an animal able to bite with great strength. **3.** The skull of *Andrewsarchus*. Its skull was 83 centimeters (33 in) long and 54 centimeters (21 in) wide. It was probably omnivorous. **4.** The skull of *Pakicetus*. This is the oldest known cetacean. It dates back to the Eocene and was found in Pakistan. Its similarity to the mesonychids, specifically *Andrewsarchus*, is clear. The teeth are not all identical like those of modern cetaceans. **5.** The skeleton of *Zygoriza*, a 5-meter (16.4-ft) long cetacean from the Eocene. The anatomical structure of this species, with its long spine, closely resembles that of modern cetaceans. **6.** A killer whale, a worthy example of today's odontocete cetaceans, which have teeth that are identical to one another. This species can grow to up to 9 meters (29.5 ft) in length.

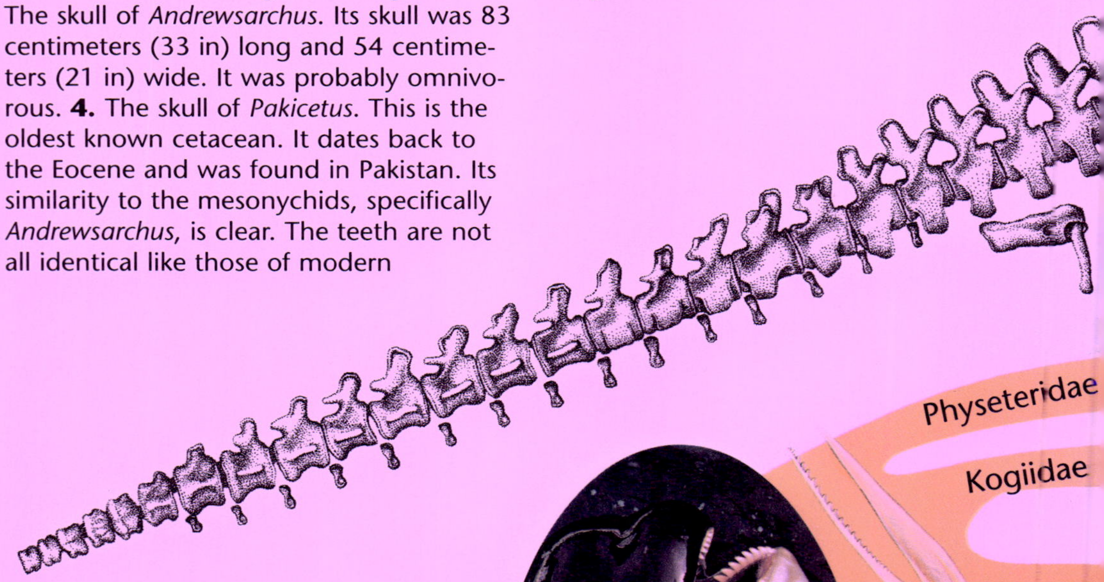

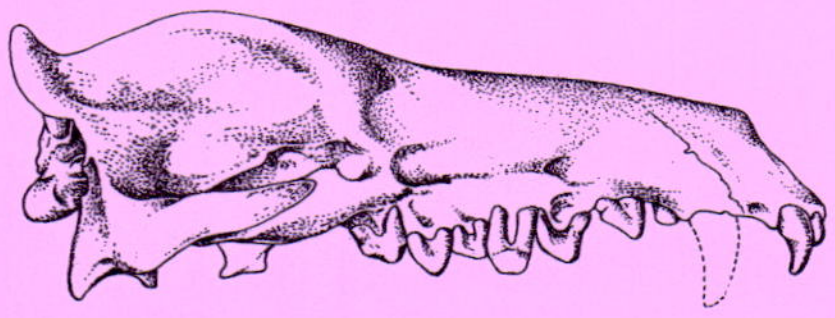

3

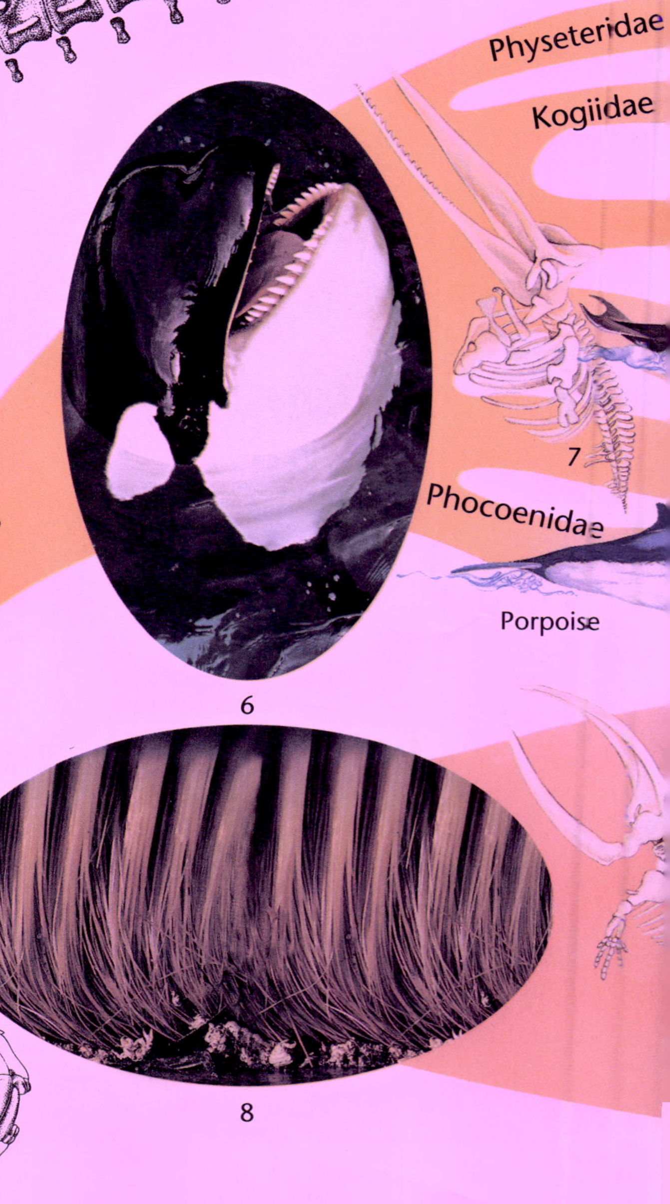

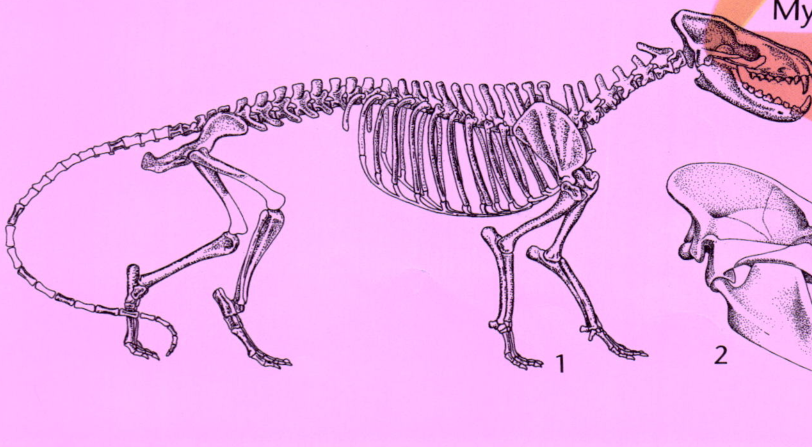

1

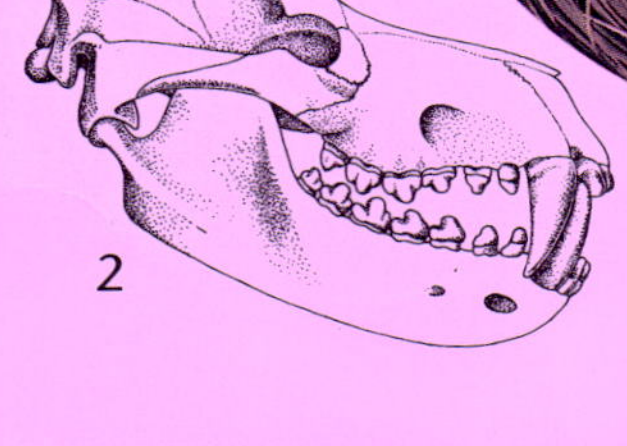

2

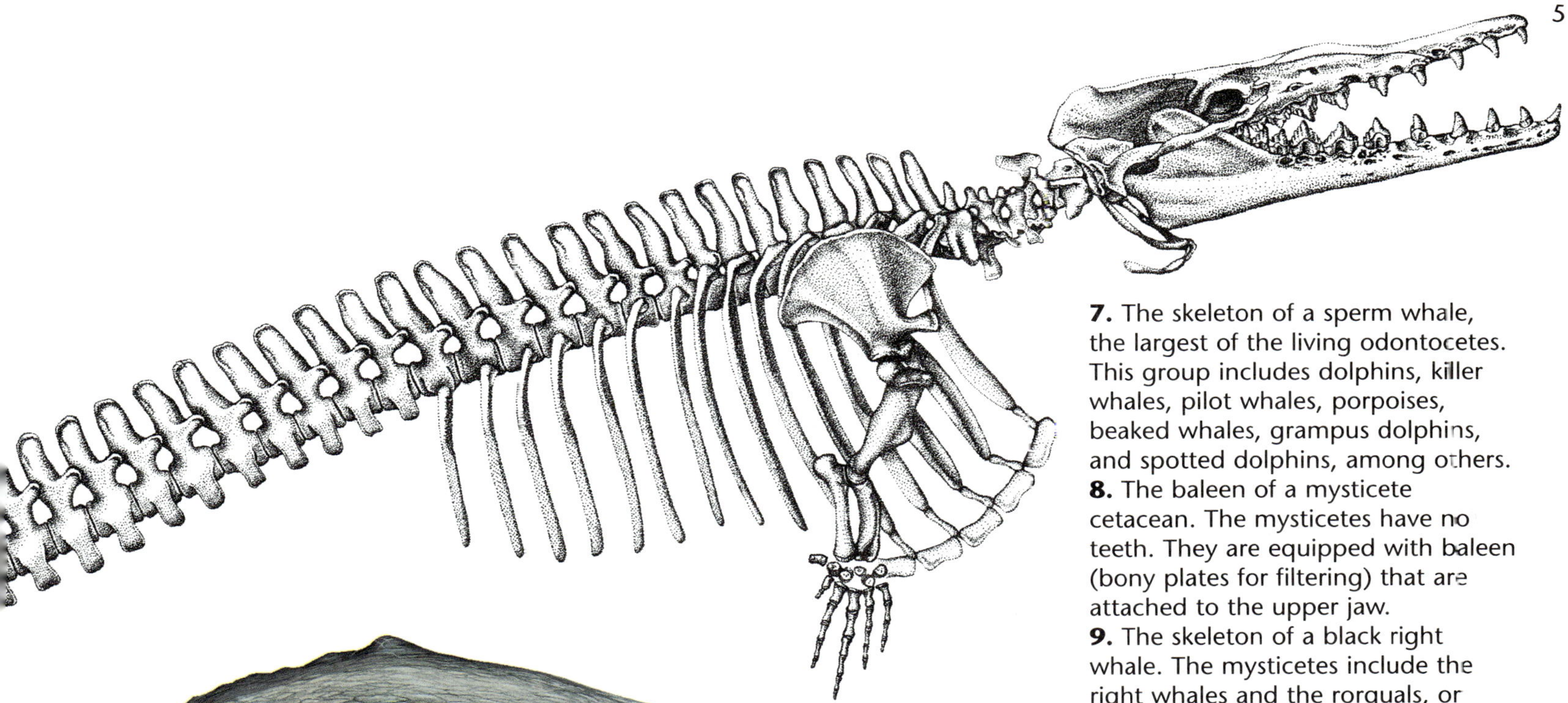

7. The skeleton of a sperm whale, the largest of the living odontocetes. This group includes dolphins, killer whales, pilot whales, porpoises, beaked whales, grampus dolphins, and spotted dolphins, among others.
8. The baleen of a mysticete cetacean. The mysticetes have no teeth. They are equipped with baleen (bony plates for filtering) that are attached to the upper jaw.
9. The skeleton of a black right whale. The mysticetes include the right whales and the rorquals, or great whales. These are the largest animals living on Earth.

Dolphins, Sperm Whales, and Killer Whales

The Adaptations of Dolphins

Starting from an ancient land mammal that walked on four legs, a fairly rapid evolutionary process led to the formation of a group of marine mammals. These animals are considered as well-adapted to the marine environment as fish. However, unlike fish, they breathe air through lungs.

1. A bottlenose dolphin, *Tursiops truncatus*, underwater. **2.** A diagram of the flow of water past the body of a swimming dolphin. The turbulence created by the movement of the tail is "washed away" as the upward movement or "power stroke" of the tail reaches its highest point. This action creates an area of low pressure, and water is drawn from the head and chest toward the tail. **3.** The dolphins' "melon" acts as a kind of acoustic lens through which the outgoing sounds can be focused. The melon is composed of fat, muscle, nasal passages, and air sacs. Incoming sounds are received by an oil-filled cavity in the lower jaw that directs them to the inner ear.

1

3

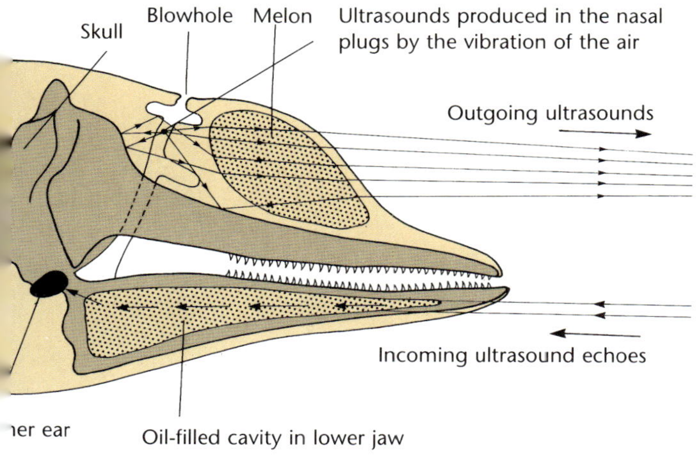

4. The lever system of bones in the middle ear of a land mammal and a dolphin. The different lengths of the levers compensate for the different sizes of sound waves traveling through air and water. In this way, the inner ear receives sound waves of equal amplitude and pressure.

4

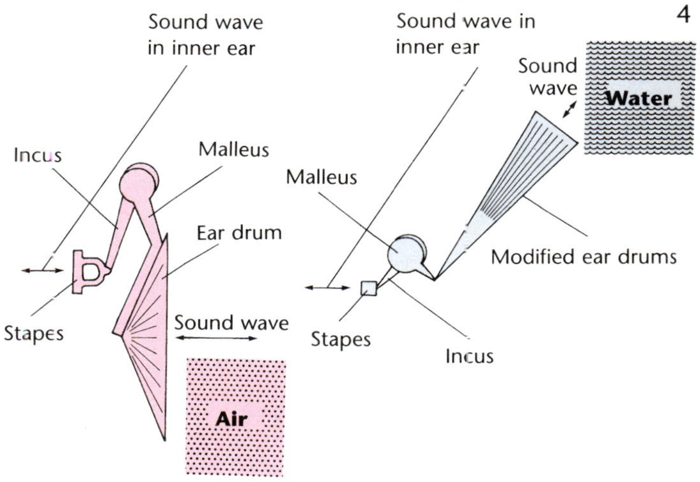

5. Dolphins and humans are the mammals with the largest neocortices and therefore are thought to be the most intelligent mammal species. Dolphins also have a large cerebellum in relation to their striking ability for orientation.

5

3

Covered by a layer of insulating fat, a typical odontocete cetacean, such as a dolphin, loses most excess heat through its flippers. These structures are well-supplied with blood. Dolphins also have unusual skin that is made up of two layers. The inner layer is spongy and flexible, as is the thin outer layer. When a dolphin swims at speeds high enough to create turbulence, its skin tends to expand or inflate. This helps the dolphin achieve a smooth flow of water around its body. As a result, the amount of energy used by the animal for swimming is decreased.

Dolphins' eyes are adapted to the refraction of light in sea water. Thus, all dolphins are short-sighted. Dolphins cannot distinguish moving objects at distances greater than 15 meters (49 ft). However, they make extraordinary use of sound. Dolphins emit a range of sounds that they can perceive using a complex inner ear. They can understand these sounds because they have one of the most extraordinary brains in the animal kingdom, humans and primates included.

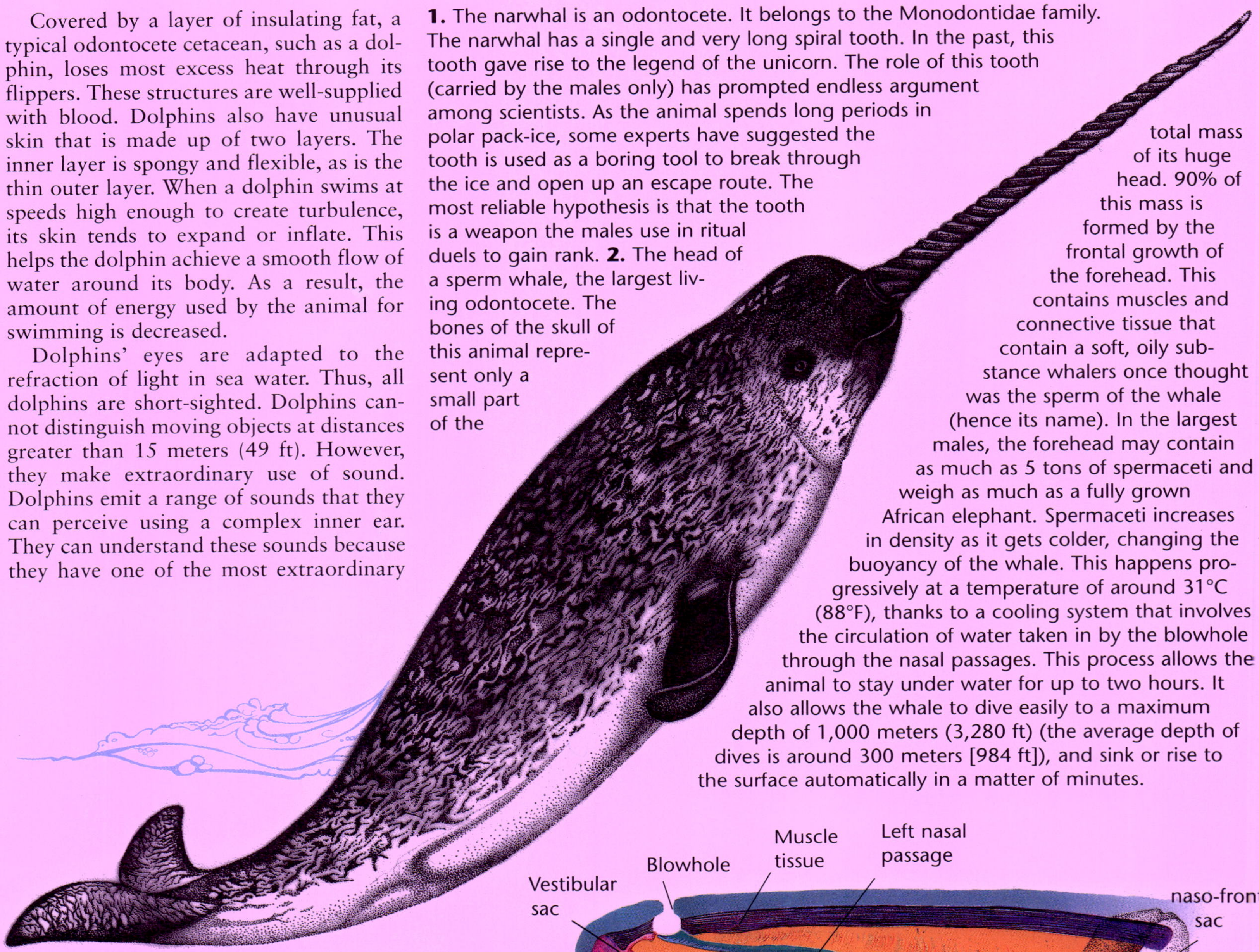

1

1. The narwhal is an odontocete. It belongs to the Monodontidae family. The narwhal has a single and very long spiral tooth. In the past, this tooth gave rise to the legend of the unicorn. The role of this tooth (carried by the males only) has prompted endless argument among scientists. As the animal spends long periods in polar pack-ice, some experts have suggested the tooth is used as a boring tool to break through the ice and open up an escape route. The most reliable hypothesis is that the tooth is a weapon the males use in ritual duels to gain rank. **2.** The head of a sperm whale, the largest living odontocete. The bones of the skull of this animal represent only a small part of the total mass of its huge head. 90% of this mass is formed by the frontal growth of the forehead. This contains muscles and connective tissue that contain a soft, oily substance whalers once thought was the sperm of the whale (hence its name). In the largest males, the forehead may contain as much as 5 tons of spermaceti and weigh as much as a fully grown African elephant. Spermaceti increases in density as it gets colder, changing the buoyancy of the whale. This happens progressively at a temperature of around 31°C (88°F), thanks to a cooling system that involves the circulation of water taken in by the blowhole through the nasal passages. This process allows the animal to stay under water for up to two hours. It also allows the whale to dive easily to a maximum depth of 1,000 meters (3,280 ft) (the average depth of dives is around 300 meters [984 ft]), and sink or rise to the surface automatically in a matter of minutes.

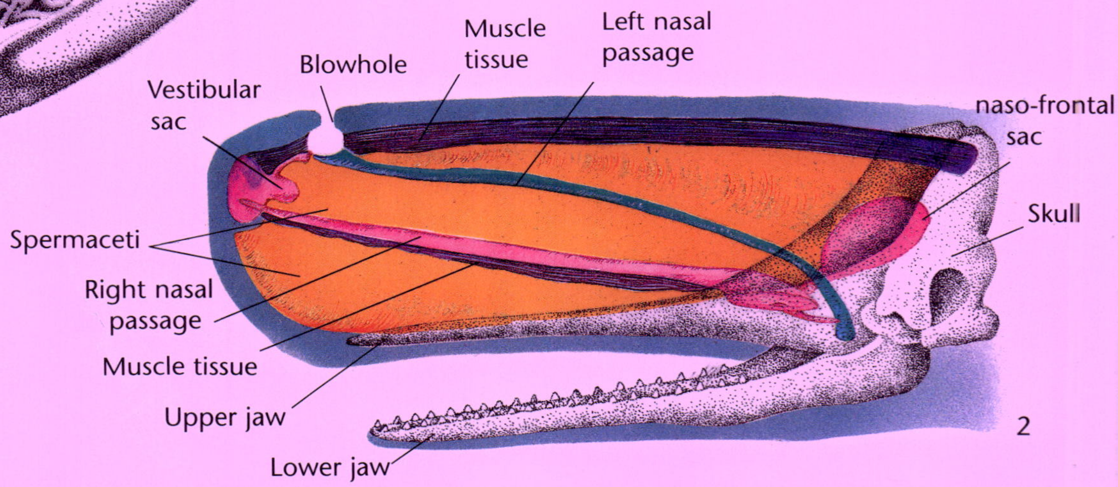

2

A Remarkable Brain

The dolphins' hearing system is more complex than that of most land mammals. Dolphins and many other members of the whale family use an **echolocation** system similar to the one used by bats. In dolphins, incoming sound is not received by the small external ear. They are instead picked up by a cavity in the lower jaw that transmits the sounds to the inner ear through two pairs of bony orifices.

The **sonar** equipment of dolphins provides these animals with a huge amount of information. Some sounds are used to gain a general idea of the presence of an object. The object can then be analyzed through different sounds and through the echoes coming from different parts of the object. The system is also an efficient way to navigate. This has been shown in several

3. A reconstruction of the attack of a large male sperm whale on a giant squid, one of its favorite foods. This was demonstrated by the finding of the remains of cephalopods up to 12 meters (39 ft) long inside the huge whale. The illustration also shows a female and two young sperm whales observing the scene in the ocean depths.

experiments in which captive dolphins were forced to find their way through mazes in the dark. Once the animals were familiar with the test, the number of mistakes they made quickly fell to zero.

As well as providing notable abilities in echolocation, the brains of dolphins can also communicate and decode complex sound messages. Several experiments have shown that dolphins can supply and receive information needed to solve problems. In these experiments, the dolphins needed to move a series of levers in a specific pattern to receive a reward. When instructions were given to one dolphin of a pair, the animal quickly communicated the information to its partner in another tank through sound. In short, dolphins have an extraordinary intellectual capacity. This capacity is suggested by the size of their brains and the number of convolutions on the brain surface (about twice as many as in human brains).

The Super-Predators of the Sea

The odontocete group—to which the dolphins belong—contains cetaceans of differing sizes. The sizes range from porpoises, which are a little over 1.5 meters (5 ft) long and weigh 50 kilograms (110 lbs), to sperm whales, which can grow to be 25 meters (82 ft) in length and weigh 50 tons. The sperm whale is a specialized predator of giant squid. This whale cannot be compared in terms of versatility or complexity of social relations with the most extraordinary of the odontocete predators, the killer whale (*Orcinus orca*), which is a form of giant dolphin.

Killer whales can grow to lengths of 6 to 9 meters (20 to 29.5 ft) and weigh over a ton (a 4-meter [13-ft] female will tip the scales at 850 kilograms [1,874 lbs]). They are easily recognizable by their characteristic black-and-white coloring. Killer whales have massive jaws armed with two rows of huge conical teeth. They often prey on seals, penguins, seabirds, cephalopods, other whales, and schools of fish. They can swim at speeds of almost 40 kilometers (25 mi) per hour. They can leap 1.5 meters (5 ft) above the water's surface for a distance of up to 13 meters (43 ft).

Killer whales hunt in groups of up to 50 members, using a variety of methods. One of the most striking hunting systems is used in the frozen polar seas to capture seals resting on the ice near open water. The killer whales dive deeply and then explode upward like missiles, breaking the ice (up to 1 meter [3.3 ft] thick) below the pinnipeds with a violent blow from their snouts. As the seals fall into the sea, the killer whales are ready to seize them. To prey on seal pups, killer whales will surf a large wave onto the beach. They swallow the prey whole with a few shark-like rhythmic movements of their jaws. Then they use the waves to float back out to sea.

In spite of their reputation, killer whales do not usually attack humans. Individuals that have been kept in captivity have been very docile. They are willing to play in the water with their keepers without ever showing aggressive intentions. This suggests that the intelligence of these animals allows them to discriminate between various situations and species and to act accordingly.

1

2
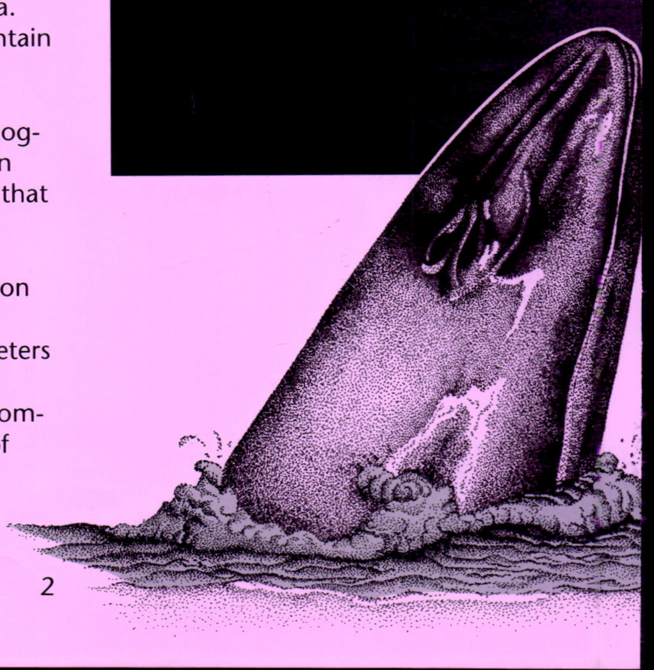

1. Killer whales swimming on the surface of the sea off Vancouver Island, Canada. The herds of killer whales generally contain a few dozen individuals and represent more or less temporary groupings of a number of families. The groups are recognizable by the "dialects" they use when vocalizing. Recent studies have shown that the groups are fairly stable. Individuals within the group maintain contact through continual sound communication based on whistles, rhythmic calls, and clicks. The tall dorsal fins (up to 1.8 meters [6 ft] high) that can be seen emerging from the water are a means of visual communication that signals the presence of adult males.

3

2. 3. Heads of blue whales (left) and killer whales while diving. The blue whale, like all mysticetes, has two blowholes. The killer whale, like all the odontocetes, has just one.
4. When whales dive, they generally empty their lungs. The oxygen supply for the time spent underwater is provided by a system that allows the gas to be stored in the blood and used little by little for the animal's metabolic needs.

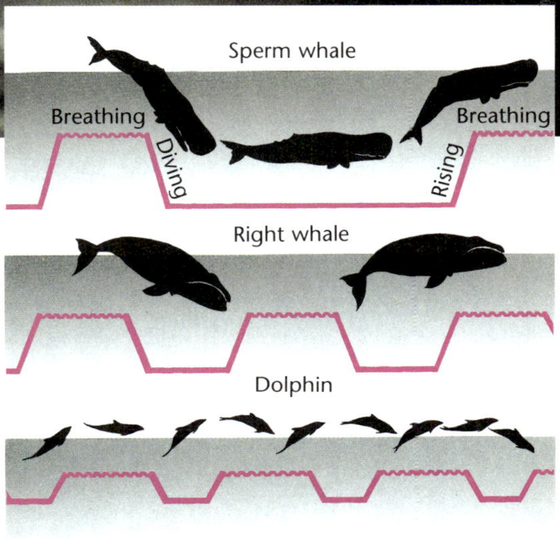

4

The Giants of the Sea

Filtering Giants

The **mysticetes**, which include right whales and **rorquals**, contain ten of the 100 living species of cetaceans. The mysticetes have completed an unusual evolutionary process in their adaptation to the marine environment. This group chose to feed on plankton, one of the most abundant resources in the sea. One result of this choice, thanks to the great availability of this new food resource, is that the mysticetes have reached very large sizes. They have also lost their teeth and evolved baleen, continuously growing horny plates that form an efficient means of filtering and capturing tiny prey. Baleen is arranged in ranks, suspended from the upper jaw. The plates are roughly triangular in shape and are fringed toward the point. They may be up to 90 centimeters (35 in) long and 30 centimeters (12 in) wide. There may be hundreds of plates suspended from a whale's jaw (the fin whale has over 900).

Another trait of the mysticetes is the presence of blowholes at the back of the head. Blowholes are nostrils that migrate

1

1. The huge tail of a humpback whale is lifted while the animal dives. This angles the body and allows a rapid descent to great depths. This action is called "fluking." **2.** The energy flow of a feeding Antarctic blue whale. Most of the food is consumed in the Antarctic region. It is either used immediately by the animal's metabolism or is kept as an energy store. **3.** The breaching, or leaping out of the water, of whales, is probably a means of communication. **4.** Phases in the breaching of a humpback whale.

2

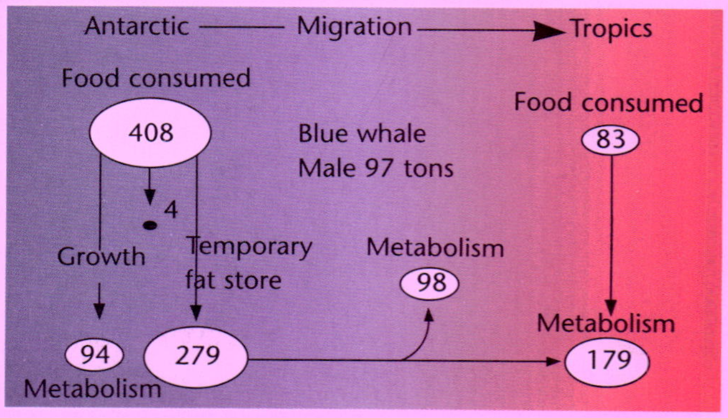

in an early embryonic stage of development (when embryos are 0.5 to 2 centimeters [0.2 to 0.8 in] long) from the tip of the snout to their new, more efficient position. The mysticetes have two blowholes, while the odontocetes have just one. This shows that the two cetacean groups began to differ long ago, and not only through the presence or absence of teeth.

A Perfect "Anti-Embolism" System

The spouting of surfacing whales is a well-known and characteristic phenomenon. It is worth trying to explain. When a diver submerges, pressure increases as the diver goes deeper. As pressure increases, more nitrogen enters the bloodstream from the diver's lungs. As the diver returns to the surface, pressure decreases. Nitrogen gas dissolved in the blood is released. If a diver rises too quickly, nitrogen may be released so rapidly that it causes many gas bubbles to form in the blood vessels. This can have fatal results.

To avoid this phenomenon, whales prepare to dive by compressing their lungs to a minimal volume. The air contained in the lungs is transferred into the trachea and the air sacs, which are reinforced with rings of cartilaginous tissue. Here, the possibility of gas entering the blood is restricted. In addition, any nitrogen that is released from the blood of a surfacing cetacean is absorbed by an oily emulsion that forms in and covers the lungs with a thin film. This emulsion, along with water

3

4

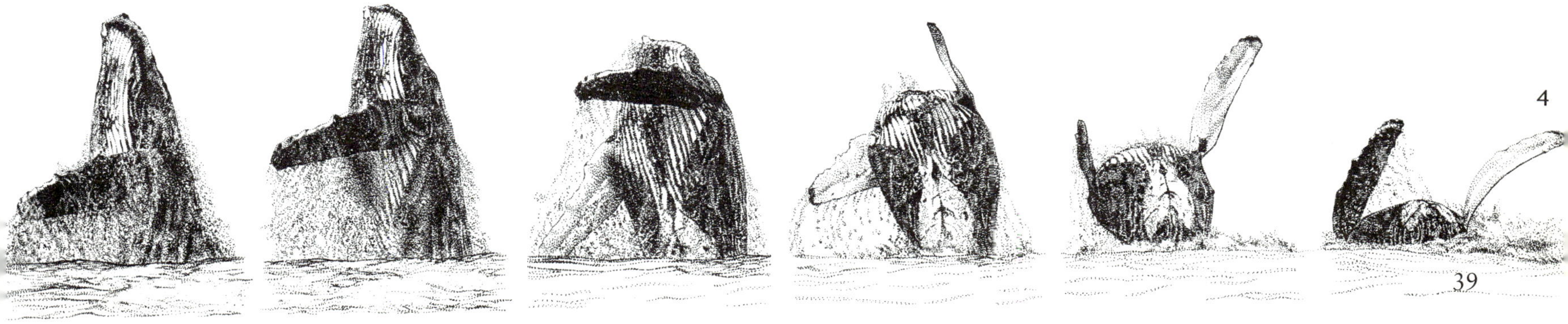

Almost all the mysticetes (this illustration shows 10 of the 13 existing species) undertake great seasonal migrations. These animals have the longest migrations in the animal kingdom. Their main source of food is krill. Krill is subject to a pronounced seasonal cycle, with strong peaks of abundance in polar waters during the summer. The mysticetes spend the warmer season in polar waters to stock up on food. They return to temperate or warm waters to give birth when krill is scarce. Due to the annual variability of local conditions and the notable versatility of many species, migration routes followed may vary greatly from one year to the next. All whales may be found in unpredictable areas. In recent years, it has been noted that the sperm whale (an odontocete that depends on giant squid for food) migrates toward the equator in autumn and returns to temperate waters in the spring.

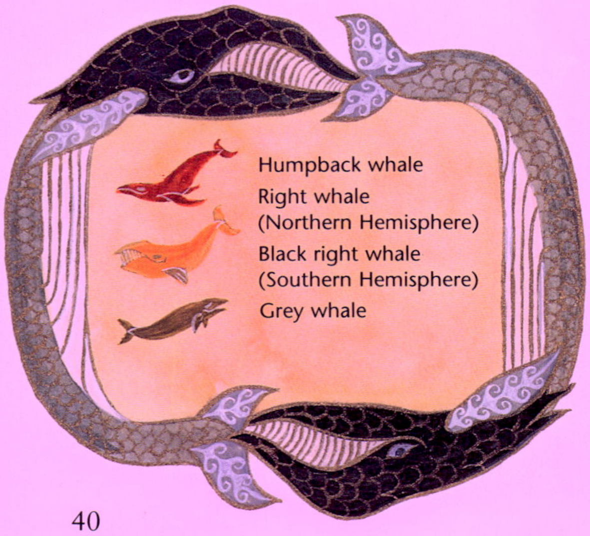

Sei whale
Balaenoptera borealis
15 meters (49 ft)
Minke whale
Balaenoptera acutorostrata
8 meters (26 ft)
Blue whale
Balaenoptera musculus
24 meters (79 ft)
75°
60°
45°
30°
15°
0°
15°
30°
45°
60°
75°
Pygmy right whale
Caperea marginata
5 meters (16 ft)
Humpback whale
Megaptera novaeangliae
14 meters (46 ft)
Greenland right whale
Balaena mysticetus
16 meters (52 ft)

vapor, is the main component of the spout whales produce as they surface.

Tiny Seasonal Food

The mysticetes have an unusual feeding system. These huge animals, up to 20 to 30 meters (66 to 98 ft) long and weighing as much as 130 tons, feed mainly on prey that is less than 5 or 6 centimeters (around 2 in) long. The prey is made up mostly of small crustaceans known as krill (*Euphasia superba*). Krill form huge schools, which is very convenient for the mysticetes while hunting.

Some whale species feed on even smaller prey, such as copepods, which are just a few millimeters long. Still others feed on fishes only a few decimeters long. Among the latter is the fin whale, a species much smaller than the blue whale. The fin whale consumes great quantities of salmonoid fish of the *Osmerus* genus.

When a blue whale swims into a school of krill, it opens its huge mouth and takes in water that is filled with crustaceans. It then compress its 2-ton tongue against its palate and expels the water through the baleen, which retains the plankton. The food is then collected by the tongue and swallowed. The stomach of a large blue whale can contain up to a ton of food.

Whales feed intensively (and perhaps exclusively) during their stay in the icy seas where food is more abundant. A blue whale can easily consume four tons of plankton each day, or 500 tons during the four months it spends in Arctic or Antarctic waters. This plankton includes a large amount of phytoplankton.

Rapid Weaning

Well-fed and stocked with reserves of fat,

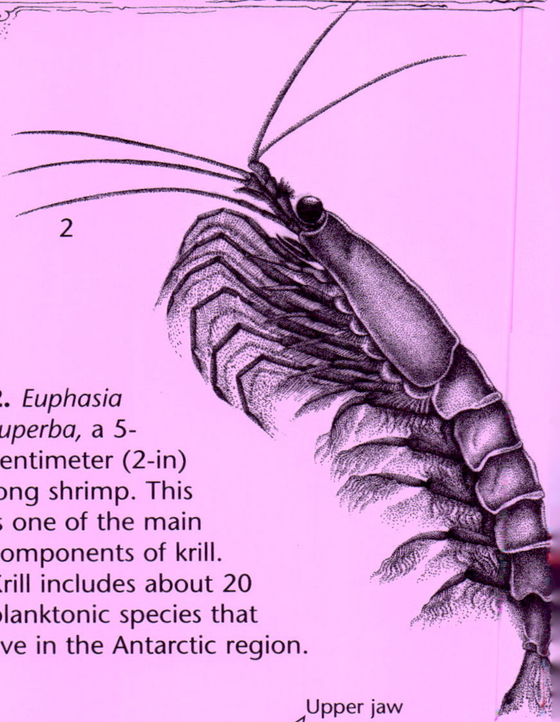

2

2. *Euphasia superba,* a 5-centimeter (2-in) long shrimp. This is one of the main components of krill. Krill includes about 20 planktonic species that live in the Antarctic region.

1

1. The gaping mouth of a minke whale, *Balaenoptera acutorostrata.* You can see the rosy palate, the creamy-white baleen fibers, and the great mass of the tongue.

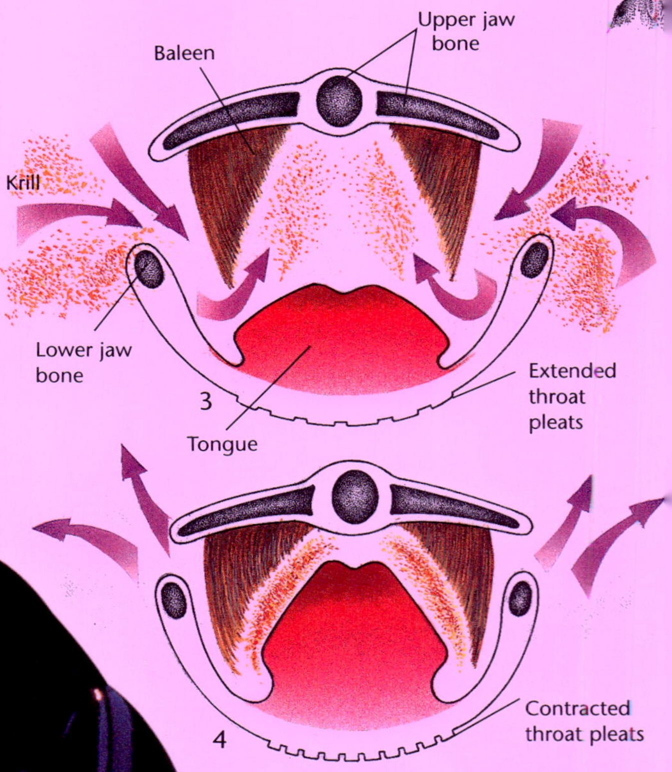

Mysticete feeding.
3. The jaws open wide, the flattened tongue retracts, the accordion-pleated skin of the throat expands like a huge sack with the weight of the water that fills it. **4.** The jaws close, the tongue thrusts toward the palate, the skin of the throat contracts, and the water is expelled. The krill is retained by the baleen filter.

right whales and rorquals migrate toward temperate and subtropical waters for the rest of the year. In these areas, the females give birth and nurse their young. During this time they will eat very little. Gradually the whales lose much of the reserves gathered during their brief stay in cold, polar waters. Newborn right whales are nursed for only six months. Blue whales are nursed for eight months. The nursing periods are surprisingly brief in relation to the sizes of the species. However, they are sufficient enough to allow the young to survive until the next migration to the polar waters. Once there, the mother and the young will find themselves in an environment rich in food resources, which is ideal for rapid weaning.

Newborn whales are nursed at the surface. This allows the young to breathe easily while feeding. After they learn to swim, they can suckle while under water. The mammary glands of cetaceans are near the genital opening. The nipples extend outward as soon as the young whale touches the surface. The milk is then actively pumped into the newborn's mouth by actions of the mother's muscles.

Young right whales and rorquals grow very quickly. A blue whale is 6 or 7 meters (20 or 23 ft) long at birth, 15 meters (49 ft) by the time it is weaned, and 21 meters (69 ft) at the age of two years. At two years, the blue whale reaches sexual maturity. The longest specimen ever recorded measured 34 meters (111.5 ft) and weighed 130 tons.

The high degree of adaptation of the blue whale to the environment is emphasized by a comparison with the sperm whale. The sperm whale is a large odontocete that is specialized in hunting large cephalopods at great depths. It does not complete seasonal migrations to polar waters. Newborn sperm whales are nursed until they can hunt alone, which can take several years.

5. In this century, the numbers of all whales have been drastically reduced. Many species face extinction due to excessive commercial hunting. The fin whale population has been reduced from 400,000 to less than 100,000. The blue whale has been reduced from 200,000 to less than 10,000. The sei whale and the humpback whale have been reduced from 100,000 to just a few thousand. Today, thanks to a ban on whaling, the populations are recovering. However, for reasons such as the consumption of krill by other species or the harvesting of the plankton by humans, this recovery is very slow.

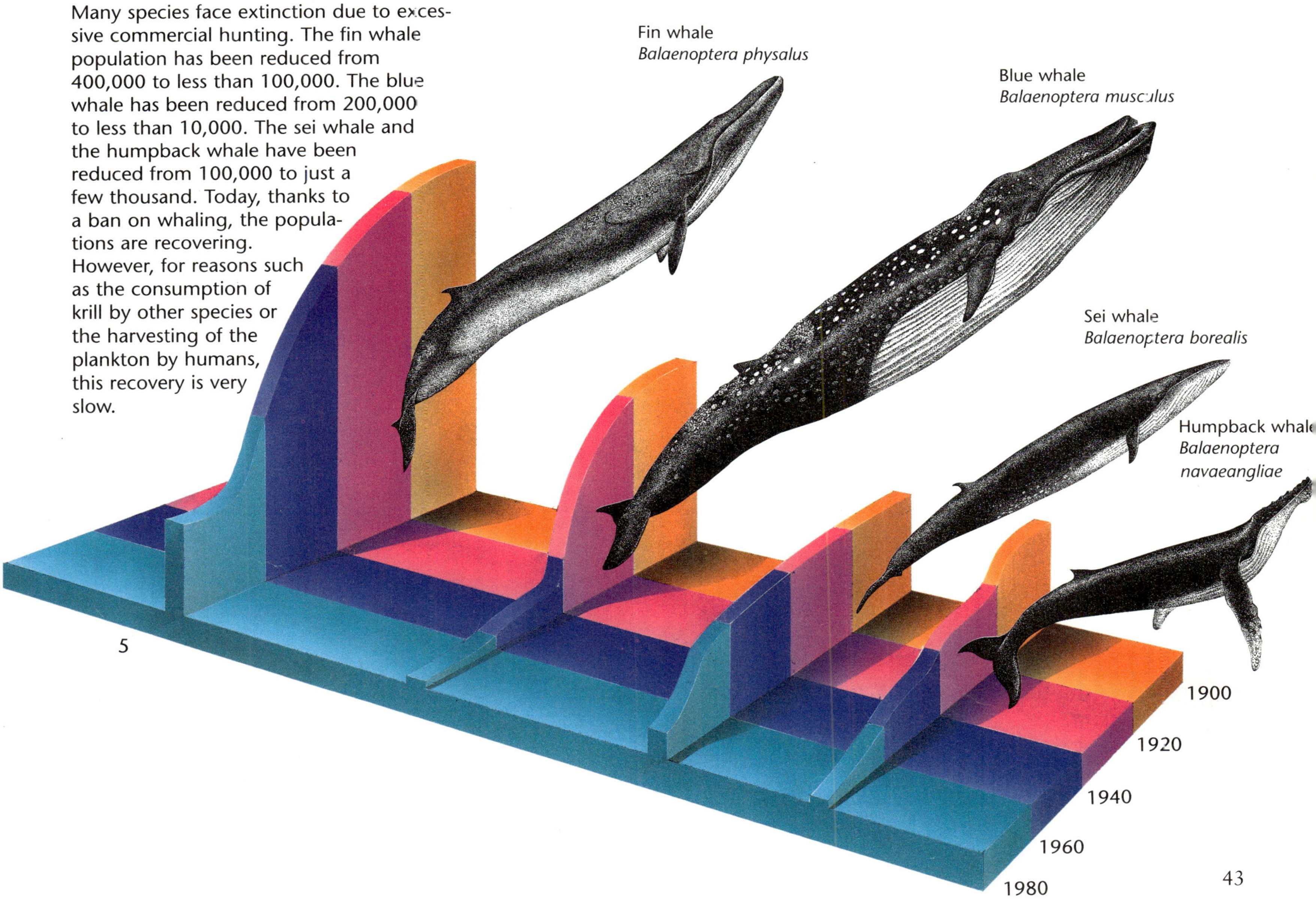

THE PAST AND THE FUTURE OF THE WHALES

1

2

1. A harpooned sperm whale is harnessed and secured against the side of a whaling ship. You can see the outrigger platform used for carving the catch, which is done with long-handled flensing knives.
2. From the platform, the flensing knives are used to cut away the head of the cetacean so the blubber can be recovered.
3. The blubber is cut into long strips and hoisted aboard the ship. The crew begins to cut up the strips. The man in the foreground chops the strips into blocks called

3

4

"horse pieces." The man behind him slices these pieces into "bible leaves."
4. A stoker throws stacks of "bible leaves" into a cauldron of boiling water with a long fork. The thin "leaves" melt rapidly. After being melted down, whale fat stays good for up to four years. Thus, whaling ships can continue with their long hunting voyages.

A great whale is not simply an extraordinary product of biological evolution. It is also a rich source of fat, meat, bone, baleen, and many other valuable substances. It was inevitable that humans would dedicate themselves to hunting whales before deciding to study and try to understand these creatures. Whaling was practiced in the Bay of Biscay as early as the 11th century. In that era, black right whales were the exclusive prey. These were much smaller than the blue whale (15-18 meters [49-59 ft] long and weighing 50 tons). They also had a lower specific weight, allowing them to float even when they were dead. This was essential for early whalers who would otherwise have lost their prey to the depths.

Toward the end of the 16th century, the black right whale became rare off the Atlantic coasts of Europe. Whalers, therefore, began to move northward, first to the coasts of Canada and then into Arctic waters. In 1712 an American whaler, Christopher Hussey was blown toward the coast of New England by a storm. There he found and killed a sperm whale. He also discovered that this species carried a great amount of oil (1-5 tons), called spermaceti, in its forehead. This substance may be used to aid the animal when diving or transmitting acoustic signals. From that moment, the sperm whale (immortalized in Herman Melville's *Moby Dick*) was hunted mercilessly to the same extent as the black right whale.

In 1864 the harpoon gun was invented. The gun could be fired directly from a ship (in earlier times the whalers relied on launching harpoons by hand from small boats). The technology was further refined with the development of floatation systems that injected air into a whale's carcass. In 1904, land stations were built where the floating corpses of the huge whales could be towed for processing. The result of all this development was a huge increase in whaling. A total of 195 whale carcasses were processed in a single land station in 1904. As many as 10,760 carcasses were processed in six stations in 1913.

These whaling stations declined in number after factory ships, fitted with all the necessary equipment for "whale-processing" at sea, were introduced. At the same time, the annual number of whales killed reached 40,000. This number remained consistent into the 1950s. Within a centu-

1

2

3

1. Hunting long-finned pilot whales, *Globicephala melaena*, in the Faroe Islands, between Iceland and Great Britain. This ancient and bloody practice relies heavily on the tendency of large groups of these whales to beach themselves. This probably results from their method of navigation in the shallows near the coastline. Inhabitants of the Faroes have long tried to provoke the disorientation and subsequent beaching of the whales by encircling them with boats, from which they throw rocks and

4

ry, the population of right whales and great whales was reduced from about 1.1 million to less than 200,000.

The world's current whale population consumes about 43 million tons of krill, 6 million tons of cephalopods, and 130 thousand tons of fish each year. These numbers were of course much larger during the last century, when there were more whales—190 million tons of krill, 14 million tons of cephalopods, and 5 million tons of fish.

Due to the smaller number of whales today, there is a notable surplus of krill. The fishing industry has begun to exploit this resource under various imaginative names. More natural beneficiaries include seabirds and pinnipeds.

To be fair, the commercial exploitation of krill by humans currently is fairly modest. It is important to note, however, that it really should not increase any further. An excessive harvest could deplete the oceans of a resource that might very well be crucial to the recovery of various whales (assuming the competition from rival consumers is not too fierce). Furthermore there is the fairly ugly possibility that humans might once again turn to whales as a source of food once the whales' numbers become "stable" again.

5

beat their oars on the surface of the water. The pilot whales are easily driven toward the shore, where their sociable habits lead them to die *en masse*. Once the pilot whales have been herded onto the beach, the fishers enter the water and kill them with a harpoon thrust in the back of the head to sever the spine.

2. The bodies of slaughtered pilot whales, before their flesh and fat are removed.

3. The same carcasses after butchering. All that remains on the beach of the lively school of whales are bones and skin. The massacre has an important role today, not so much in the economy of the remote islands as in the local culture that has traditional ties with this ancient hunting practice. It is called the "grind" in the local dialect. Written records of the practice date back to 1584.

4. A stranded pilot whale in Maryland. The stranding of pilot whales is fairly common. It can easily take place without the intervention of hunters.

5. A group of volunteers attempts to push a stranded pilot whale back out to sea. For the most part, pilot whales, like all the other Delphinidae species, are loved and respected creatures. Today they are generally protected rather than persecuted.

Glossary

(Cross references appear in *italics*)

adaptation Term used to describe the adjustment of the traits of an organism to its environment. Long-term adaptation is genetic and occurs in a population through natural selection. An individual may adapt through physiological, or behavioral, changes through learning.

Antarctic Area in the Southern Hemisphere between the South Pole and Antarctic Circle.

Arctic Area in the Northern Hemisphere between the North Pole and the Arctic Circle.

auk Member of a family of *pelagic* seabirds similar in appearance to a *penguin*. Auks have black-and-white plumage, short wings, and often laterally compressed beaks. Auks nest in colonies in tunnels, cracks, or niches in the rock, except for *Brachyramphus marmoratus*, which builds its nests in the branches of conifers.

austral Relating to the Southern Hemisphere, from the South Pole to the Equator.

Balaenidae *Mysticete cetaceans* with huge mouths that have *baleen,* but lack pleated throats. A typical example is the bowhead, or Greenland right, whale (*Balaena mysticetus*).

baleen Continuously growing horny plates located in the mouths of *mysticete cetaceans*. Baleen provides an efficient means of filtering and capturing tiny prey. Arranged in rows suspended from the upper jaw, the roughly triangular baleen is fringed toward the point and can be up to 90 centimeters (36 in) long and 30 centimeters (12 in) wide.

bird Member of a class (Aves) of *vertebrates* having constant body temperatures. Birds evolved in the late *Jurassic* Period from small dinosaurs and include many orders and more than 9,000 species. Hundreds of birds belong to about 12 families that have adapted to the marine environment to varying degrees.

boreal Of the Northern Hemisphere, from the North Pole to the Equator.

bos'n bird Member of a family of medium-sized seabirds that are mainly white in color, with strong red or yellow beaks and long tails. Highly aerial and *pelagic*, they feed mainly on fishes caught by diving.

carapace The domed back (dorsal) part of the shell of turtles and tortoises.

Carboniferous Period in the Paleozoic Era from about 360 million to about 290 million years ago. Characterized by the abundance of fossil carbon deposits and rich terrestrial and aquatic animals, including the first insects, many amphibians, and the first *reptiles*.

cetacean Marine *mammal* in the family commonly called the whales. This group includes dolphins and porposies as well as killer whales (orcas), humpbacks, and fin whales, among others.

Colubrids Any member of the large snake family Colubridae. Most are harmless and have small, solid, non-channeled teeth unsuitable for injecting *venom* (aglyphs). In some cases, they have *venom* fangs set well back in the jaw (opistoglyphs). The right-hand lung is well-developed, while the left lung is rudimentary or absent.

cormorants Family of aquatic *birds* that are generally dark in color, have a long beak with a hooked tip, relatively short wings, and a long tail. They usually fly just above the water's surface and dive to catch fishes. One of about 30 existing species, the Galápagos cormorant (*Phalacrocorax harrisi*) is flightless.

Cretaceous Period in the Mesozoic Era from about 140 million to about 65 million years ago. Characterized by the thriving of many species of *reptiles* (including the first sea turtles) and the current presence of fairly undiversified *birds* and *mammals.* Toward the end of the period, the dinosaurs, *ichthyosaurs*, *plesiosaurs*, *nothosaurs*, and ammonites went extinct leading to a rise of the new animals of the Tertiary Period.

Devonian Period in the *Paleozoic* Era from about 410 million to 360 million years ago. The *vertebrates* evolved during this period along with armored fishes lacking jaws and bony and cartilaginous fishes found in fresh water and estuaries. The invasion of dry land that began in the Silurian was by now complete for both plants and animals. The amphibians developed from a group of bony fishes, the *Crossopterygii*.

echolocation The locating of an object through a system that uses reflected sound. It is used by bats and *cetaceans*.

Eocene Epoch in the Tertiary Period from about 55 to 37 million years ago. The Earth was populated by a variety of *mammals* and *birds. Penguins* developed from an ancient group similar to present-day loons.

fossil Remains of an organism that lived in the past, usually before the last Ice Age (at least 10,000 years ago). Fossils include skeletons, footprints, tracks, marks, perforations, and replacement forms in silica, chalk or other mineral substances.

frigates Family of large seabirds with elongated bodies, long narrow wings, deeply forked tails, and long hooked beaks. They are powerful fliers but lack watertight plumage and therefore never float on the surface of the water. They try to feed by snatching the prey of other seabirds while in flight.

gannets Family of seabirds having long conical beaks lacking nostrils, long pointed wings, wedge-shaped tails, and webbed feet that are often brightly colored. They feed on fishes caught by diving and nest in dense colonies on small islands or rocky coastal cliffs. There are nine species distributed in all the seas of the world.

ichthyosaurs Order of extinct *reptiles* that appeared in the *Triassic* and became extinct in the *Cretaceous.* They had shapes similar to modern dolphins.

invertebrates Animals that lack backbones.

Jurassic Period in the Mesozoic Era from about 210 to about 140 million years ago. The dinosaurs were thriving, and many small tree-dwelling species developed and gave rise to the first *birds*. The *ichthyosaurs* appeared in the seas, deriving from ancient land *reptiles* of the extinct *Parapsidae* group.

mammals Class of land *vertebrates* having constant body temperatures, hair, and mammary glands. They originated toward the end of the *Triassic* (210 million years ago) and became highly diversified early in the Tertiary, around 150 million years later. The only mammal order that has adapted to life in the open sea is the *cetaceans*. Some carnivores (otters and pinnipeds) and sirenians (dugongs and manatees) are less adapted to an aquatic environment and inhabit coastal areas.

Miocene Period in the Tertiary Era from about 23 to about 5 million years ago. Modern sea snakes (Hydrophiidae) originated during this period.

mysticetes Suborder of *cetaceans* that includes the true whales. They lack teeth but have *baleen*.

nothosaurs Suborder of marine *reptiles* that had its period of maximum development in the *Triassic* Period. These reptiles had long necks and front limbs modified for swimming. Early in the *Jurassic* they disappeared and were replaced by the *plesiosaurs*.

odontocetes Suborder of *cetaceans* that includes all species having teeth (dolphins, killer whales, porpoises, narwhals, beaked whales, and grampus dolphins).

Oligocene Epoch in the Tertiary Period from about 37 to about 23 million years ago. The *cetaceans* developed during this period from ancient land *mammals* that are now extinct (mesonychids) but were similar to modern carnivores.

Paleocene Epoch in the Tertiary Period from about 65 to about 55 million years ago. Characterized by the adaptive development of *mammals* that continue to differentiate and occupy many ecological niches left by the extinct reptiles of the *Cretaceous*.

pelagic Term to describe organisms that live in open seas (from the Greek *pelagos*).

Pelecanidae Family of large *birds* having long, broad wings, short tails, and large beaks with extendible throat sacs. They live in estuaries and inland water, feed on fishes and crustaceans, and nest in colonies on the ground or in large trees.

penguins Family of large or medium-sized seabirds having very small feathers covering the entire surface area of their bodies. Some species have crests over their eyes. All are flightless but are excellent swimmers because their wings have evolved into flippers. They feed on fishes caught under water and nest on land.

plastron The bottom (ventral) part of a turtle's shell.

plesiosaurs Suborder of aquatic reptiles that appear among the *fossils* of the Upper *Triassic* and became common in deposits from the *Jurassic* and the *Cretaceous*. They were equipped with large, powerful bodies. Some had short necks, others had swan-like necks, and still others had long, serpentine necks. Active carnivores, some species reached lengths of up to 15 meters (49 ft).

Procellariidae Order of *birds* that includes albatrosses, shearwaters, fulmars, petrels, and storm petrels. They have tubular nostrils and secrete an oil with a strong fishy smell that can be regurgitated as a form of defense. They are all *pelagic* and feed on fishes, plankton, crustaceans, and other *invertebrates*. They nest on small islands or remote rocky coasts.

reptiles Class of land *vertebrates* with a variable body temperature that developed from amphibians in the *Carboniferous*. They have bodies covered with scales or bony plates. The class includes the orders of the Chelonia (turtles), crocodiles, Squamata (lizards and snakes), and the Rhynchocephalia (the tuatara of New Zealand, a living *fossil* that has survived almost unchanged since the *Jurassic*). A few species (5 turtles and around 50 snakes) have adapted to life in the open sea, while just two, the marine iguana and the marine crocodile, live along the coast of particular geographical areas.

rorquals *Mysticetes* with huge mouths, large *baleen*, and pleated throats. The last feature allows the throat to be expanded so a large amount of water can be taken in and filtered. A typical example is the blue whale (*Balaenoptera musculus*), the largest living cetacean—it can reach up to 33 meters (108 ft) in length and weigh up to 130 tons—and, as far as we know, the largest animal on Earth.

Sauropterygia Extinct order of reptiles that includes the suborders of the *nothosaurs* and the *plesiosaurs*.

scales Epidermal (skin) formations typical of reptiles and fish. Larger scales, symmetrically arranged on the head, are known as plaques. Those on the belly and lower part of the tail are called plates. In turtles, the shells are composed of bony plates covered with horny scales.

sound waves Longitudinal waves originating from organs that can vibrate and transmit vibrations to an elastic medium such as air or water. Violin strings, human vocal cords, the syrinx of songbirds, and the chirping organs of insects are all examples.

sonar Instrument or system capable of locating an object by using reflected sound.

Spheniscidae Scientific name for the *penguin* family.

Triassic First and oldest period in the Mesozoic Era, from about 250 to about 210 million years ago. Well-differentiated reptiles populated the Earth, and the first *mammals* evolved—small nocturnal animals similar to the modern insectivores. The extinct group of Parapsidae reptiles saw the *adaptation* to an aquatic lifestyle of the *Sauropterygia* order that gave rise to the suborders *nothosaurs* and *plesiosaurs*.

ultrasounds Sounds of frequencies that are too high to be heard by the human ear.

venom In serpents, the liquid secreted by modified salivary glands that contain mostly protein in the form of enzymes. It can provoke serious effects when injected into the bloodstream of an animal (via fangs). The venoms include neurotoxins that act on the nervous system, hemorrhagic agents that destroy the lining of the blood vessels, coagulants that produce blood clots in the blood vessels, hemolysins that destroy red blood cells, substances that destroy other cells, anticoagulants, anti-bactericides, and predigestive ferments. These substances form various compounds used by the different species of venomous snakes, and the effects produced vary widely.

vertebrates Animals that have backbones.

Further Reading

Bailey, Marilyn. *Evolution: Opposing Viewpoints.* (Great Mysteries Series). Greenhaven, 1990
Bramwell, Martyn. *The Oceans* (revised edition). Watts, 1994
Dal Sasso, Cristano. *Animals.* (Beginnings Series). Raintree Steck-Vaughn, 1995
Davis, Lloyd S. *Penguin: A Season in the Life of the Adelie Penguin.* Harcourt Brace, 1994
Dow, Lesley. *Whales.* (Great Creatures of the World Series). Facts on File, 1990
Editors, Raintree Steck-Vaughn. *The Raintree Steck-Vaughn Illustrated Science Encyclopedia* (1997 edition). (24 volumes). Raintree Steck-Vaughn, 1997
Emery, Richard. *Great Moments in Evolution: A Trilogy.* Eon Entertainment, 1995
Green, Carl R. and Sandford, William R. *The Bottlenose Dolphin.* (Wildlife Habits & Habitats Series). Silver Burdett, 1987
Hoyt, Erich. *Riding with the Dolphins: The Equinox Guide to Dolphins & Porpoises.* Firefly Books, 1992
Klobas, John. *Life Cycle of the Pacific Gray Whale.* (Life Cycles Series). Heian International, 1993
Lambert, David. *The Pacific Ocean.* (Seas and Oceans Series). Raintree Steck-Vaughn, 1996
Lambert, David and McConnell, Anita. *Seas and Oceans.* (World of Science Series). Facts on File, 1985
Markle, Sandra. *Pioneering Ocean Depths.* Simon and Schuster, 1994
Mattson, Robert A. *The Living Ocean.* (Living World Series). Enslow, 1991
May, John, ed. *The Greenpeace Book of Dolphins.* Sterling, 1992
Morton, Alexandra. *In the Company of Whales: From the Diary of a Whale Watcher.* Orca Books, 1993
Neal, Philip. *The Oceans.* (Conservation 2000 Series). Trafalgar, 1993
Pringle, Laurence. *Dolphin Man: Exploring the World of Dolphins.* Simon and Schuster, 1995
Twist, Clint. *Charles Darwin: On the Trail of Evolution.* (Beyond the Horizon Series). Raintree Steck-Vaughn, 1994
Waterlow, Julia. *The Atlantic Ocean.* (Seas and Oceans Series). Raintree Steck-Vaughn, 1996

PICTURE CREDITS

Photographs
GIANNI ARCUDI, Bergamo: 2-3, 14-15, 23, 30 (6), 32-33. ARDEA LTD., London (FRANCOIS GOHIER): 30 (8). GERALD CUBITT: 20 [published in *Wild Malaysia*, by G. Cubitt and J. Payne, New Holland Publishers, London, 1990]. EDITORIALE JACA BOOK, Milano (RENATO MASSA): 25; (CARLO SCOTTI): 10-11 (1). GRANATA PRESS, Milano (GORDON E. CALDWELL): 47 (5); (D. CURREY): 46 (1); (CHRIS HUSS): 36-37; (SIPA PRESS): 38. GRAZIA NERI, Milano (PETER CARRETTE): 47 (4); (JM CHARLES): 28-29; (MARTTI LINTUNEN): 46 (2, 3); (OSOLINSKI): 28; (RON SANFORD): 39; (MARK VOTIER): 42. GIOVANNI PINNA, Milano: 10, 11 (5).

Color plates and drawings
EDITORIALE JACA BOOK, Milano (LUISA CITTONE): 30-31 (5); (ANNAPAOLA DEL NEVO): 15, 16-17; (MARIA ELENA GONANO): 8, 12-13, 18-19, 26-27, 40-41, 44-45; (LORENZO ORLANDI): 11; (GIULIA RE): 20-21, 22-23, 30-31, 33 (5), 34 (2), 35, 43; (CRISTINA TRALLI): 32 (2), 34 (1), 36-37 (2, 4), 42 (2, 3, 4).

The art on page 30 (1, 2, 3, and 4) is from *Vertebrate Paleontology and Evolution*, by Robert L. Carroll, published by W. H. Freeman and Company, New York, 1988. The art on pages 32-33 (2, 3, and 4), page 38 (2), and page 42 (3, 4) is from *Whales of the World*, by Nigel Bonner, published by Blandford Press, London, 1989.

INDEX

Note: page numbers in *italics* indicate illustrations